CHEMICAL TECHNOLOGY HANDBOOK

CHEMICAL TECHNOLOGY HANDBOOK

Guidebook for
Industrial Chemical Technologists
and Technicians

by the
Writing Team for the Chemical Technician Curriculum Project

edited by
Robert L. Pecsok, *Project Director*
Kenneth Chapman and Wade H. Ponder, *Associate Project Directors*

AMERICAN CHEMICAL SOCIETY
WASHINGTON, D. C.   1975

Library of Congress CIP Data

Chemical Technician Curriculum Project. Writing Team.
Chemical technology handbook.

Bibliography: p.
Includes index.

1. Chemistry, Technical—Handbooks, manuals, etc.
I. Pecsok, Robert L. II. Chapman, Kenneth. III.
Ponder, Wade H. IV. Title.

TP151.C575   1975   660   75-22497
ISBN 0-8412-0242-7

Contents

Contributing Writers

Jack T. Ballinger
W. Robert Barnard
Nathaniel Brenner
Clark E. Bricker
C. Herbert Bryce
Samuel J. Castleberry
Allan R. Croft
Lawrence E. Dalen
Harry G. Hajian
Charles F. Hammer
Donald G. Hicks
Robert A. Hofstader
Peter C. Jurs
Donald A. Keyworth
William F. Kieffer
Charles M. Knobler
Robert D. Krienke
J. J. Lagowski
Aubrey L. McClellan
William H. McFadden
Clifton E. Meloan
John H. Paxton
Wade H. Ponder
J. Howard Purnell
William Royal
Fred W. Schmitz
H. Vernon Seklemian
David A. Shirley
Jack Sosinsky
James H. Thomas
William J. Wasserman
Thelma Williams

Preface

Chemistry is a demanding profession which requires accurate, efficient, and skillful work in the laboratory and pilot plant. Anyone who works in a chemical laboratory will find *The Chemical Technology Handbook* to be a valuable guide and reference work. Information on all aspects of laboratory work—some of which is often difficult to find—has been collected in one volume.

Whether you use this *Handbook* as a text or as a reference, you will be better able to research assignments, design and build the necessary apparatus, work in the laboratory with maximum safety and efficiency, do calculations easily, present results meaningfully, and suggest further work. Increasing your ability to participate in the entire research or production process will make your job more interesting and fulfilling no matter what your specific assignment is.

The Chemical Technology Handbook is the revised version, with additional chapters, of the guidebook from the series, "Modern Chemical Technology," which was developed by the Chemical Technician Curriculum Project (ChemTeC). The text series was developed by the American Chemical Society under a grant from the National Science Foundation.

First Aid

"HELP! HELP!" If you heard that cry from a co-worker would you be ready with some basic knowledge of first aid? If you found an unconscious co-worker how would you react? If you were the unconscious body, your life might depend on whether or not the first co-worker to get to you was prepared to do the right thing promptly. To the victim first aid is of vital importance.

Knowing the right things to do in an emergency can also be important off the job. What follows here is not a first aid course but a brief review of some possible situations that may confront a laboratory or plant technician. Practically all of these situations need immediate attention and some demand immediate action. There will be no time to refer to books in the library; you must know what to do, and you must do it immediately.

Emergency Treatment

1. Avoid panic or excitement.
2. Immediately send for a physician.
3. If immediate action is necessary to save a life, do it *without delay:*
   first, control bleeding
   second, give artificial respiration
   then treat physical shock
4. Try to determine if there are other injuries.

5. Never move a badly injured person unless it is necessary to get him to fresh air or protect him from further injury.
6. Send for the police, an ambulance, or the fire department if needed.
7. After treating for bleeding, breathing stoppage, or physical shock, treat open wounds, burns, and fractures.
8. If a physician cannot be summoned, take the victim where medical help is available.

This manual is not the place for an extended description of symptoms and treatments. An excellent wallet-sized booklet, "Pocket Guide to First Aid," is available from offices of the National Safety Council, and we strongly urge you to get one and carry it with you. More detailed books and manuals are available from the American Red Cross, the Bureau of Mines in the U.S. Department of the Interior, and any library.

Several of the most important steps in providing aid are briefly discussed below. Laboratory personnel should be able to take these important steps without wasting time.

Heavy Bleeding

For treatment of a wound, use a pad of the cleanest material available and apply pressure *directly* on the wound.

As bleeding slows or stops, tie the pad firmly in place with a strong bandage or with cloth strips. For pressure points, refer to a first aid booklet. (Warning: Use a tourniquet only if everything else fails.)

Artificial Respiration

Although several methods of artificial respiration can be used, the mouth-to-mouth method is preferred. The following instructions apply to this method.

1. Examine the victim's mouth for foreign matter (mucus, food, sand, tobacco, loose dentures, etc.). If any is present, turn his head to one side and remove it with your fingers or a cloth wrapped around your fingers.

2. Lift the victim's neck and place a folded coat, blanket, etc. under his shoulders. Tilt his head back as far as possible.

3. Grasp the jaw with thumb in one side of the mouth and pull it forward. Maintain this position to keep the air passage open.

4. Pinch victim's nostrils shut, take a deep breath, and place your mouth over his mouth and your thumb, creating a tight seal; or close the victim's mouth, take a deep breath, and place your mouth over his nose. Blow into victim's mouth or nose until you see his chest rise. For an infant, breathe through both nose and mouth, with your thumb in mouth.

5. Remove your mouth and listen for out-flow of air. For an adult, inflate lungs at a rate of about 12 times per minute. For a child, inflate lungs up to 20 times per minute, using relatively shallow breaths.

If the first few attempts to inflate the lungs are unsuccessful, turn the victim on his side and administer several sharp blows between the shoulders in an attempt to dislodge the obstruction.

NOTE: A handkerchief placed over the victim's mouth or nose prevents the need of direct contact. This does not greatly affect the passage of air.

Physical Shock

Physical shock is very dangerous and, if not recognized and treated promptly, could cause death. It may accompany severe injuries, pain, burns, loss of blood, serious fright, or poisoning. Its action may be delayed by minutes or hours.

Burns

Burns need protection from contamination and require medical attention. The victim needs to be protected from shock, and his pain should be reduced if

Table I. Symptoms and First Aid for Physical Shock

Symptoms	*First Aid*
1. Pale face	1. Lay victim down, make comfortable, with head level or slightly downhill
2. Anxious, dull look	
3. Eyelids droop, dull eyes, large pupils	2. Cleanse mouth, turn head to clear vomit, give fresh air, wrap over and under for warmth
4. Partially or totally unconscious	
5. Cold, clammy skin, subnormal temperature	3. Apply stimulants by smell if unconscious, by mouth if conscious (aromatic spirits of ammonia, coffee, tea, NO ALCOHOL), have him sip the hot liquid slowly to regain body warmth and thus blood pressure and organ activity
6. Pulse very weak and rapid, breathing very shallow	
7. Acts stupid or uninterested	
8. Cloudy vision, dizzy, thirsty	4. Keep down and quiet
9. Nausea and vomiting	5. Call physician and look for other injuries
10. Drop in blood pressure	

possible. For extensive chemical burns, the victim should immediately get to a safety shower so that large volumes of water can be used for washing. Clothing that may have become soaked with the chemical should be removed immediately. Get the victim to lie down, then treat for shock.

Thermal (heat) burns are best treated by applying clean, cold water (ice water if possible) to the affected area until a physician can be summoned. This stops the intense pain and reduces the severity of the burn. If this cannot be done, cover the burned area *very lightly* with the cleanest cloth available. Get a physician to come at once. If this is impossible, take the victim to a physician or to a hospital.

CAUTIONS: Watch carefully for symptoms of shock and treat it if they appear. Do not apply oils, greases, ointments, or baking soda to extensive burns.

Fractures

In case of fracture, get medical help for the patient. Watch for symptoms of shock and treat the victim if they appear. Never move the injured person without applying a splint to the fracture. If the fracture is compound (the skin is broken), control the bleeding but handle the broken limb with the greatest care. If you do not know how to apply a splint, look in a first aid booklet or find someone who knows how splints should be applied. Fractures do not usually need the instant attention that bleeding and arrested breathing demand.

Poisons

Poisons are discussed partially in Chapter 5, "Toxic Chemicals."

Call a physician at once. If the victim is conscious, give him large quantities of water or milk. Whether or not vomiting should be induced depends upon the poison. Do not induce vomiting if there is severe pain or burning in the victim's mouth and throat or if he is known to have swallowed a petroleum product, alkali, strong cleaning compound, iodine, washing soda, acids, ammonia, or bleach. If vomiting is safe, try pressing your finger on the extreme back of the victim's tongue where the throat begins. Some liquids can also be used to induce vomiting. Place him face down with his head low to keep vomitus out of his lungs. If he has trouble breathing give artificial respiration. Watch for symptoms of shock.

Fainting

Fainting is a mild form of shock which may develop into serious shock if left untreated. Weakness, dizziness, and rapid, weak pulse are usually followed by brief unconsciousness. The victim becomes pale, perspires, and his breathing is shallow. If a fainting attack is caught early enough, the victim should sit with his head between his knees. If the fainting attack has progressed further, have the victim lie down and keep his head low. Tight clothing such as shirt collars and ties must be loosened. If the victim is cold cover him and keep him quiet for half an hour.

Heat Stroke and Heat Exhaustion

Heat stroke and heat exhaustion do not occur frequently in the laboratory, but they may occur in a plant or outdoors. The conditions are quite different and both are serious. The victim of heat stroke is very hot and dry, and his temperature must be immediately reduced. The victim of heat exhaustion feels chilly and should be treated as if in shock. It would be wise to obtain more detail on these two conditions from a first aid reference. There are various degrees of shock, all of which are very dangerous or may become dangerous if not properly treated.

In nearly all well-organized laboratories or plants, physicians, nurses, or trained first aid specialists are readily available. If first aid is needed and time permits, the technician should not use first aid procedures himself. The exception occurs when *immediate* help must be given to the victim while waiting for the trained specialist to arrive. Your help alone may be the difference between life and death for the victim.

Electrical Shock

Treatment for severe electrical shock and how to rescue a person in contact with live electricity are described in Chapter 8, "Electrical Hazards."

2

Good Practices in the Chemical Laboratory

The term "chemicals" includes some harmless substances. It also includes many other substances such as explosives, liquid, solid, and gaseous poisons, flammable materials of all kinds, and materials like strong acids which burn the skin. Chemical workers constantly use these materials, yet the number of accidents occurring in chemical laboratories is quite low. A chemical laboratory is actually safer than the average home because laboratories and their contents have been carefully designed for safe operation, and the scientists and technicians who work in them have been trained to consider safety an important aspect of their work. Safe practices may become routine to a well-trained chemical technician, but they must always be applied consciously and regularly. Learning laboratory safety habits is something like learning to drive an automobile safely—at the beginning it takes deliberate effort, but when properly learned it becomes almost second nature. As in driving, safe practice in the laboratory requires a little effort, but the important result is prevention of injury or damage.

Laboratory Safety Practices

Your laboratory safety practices are just as important to your employer or your school as they are to you since your personal safety, the safety of your fellow workers, and the protection of property and equipment are important to them. Therefore, the laboratory will be provided with equipment designed to help prevent accidents and to prevent or reduce injury or damage should accidents occur.

The first safety rule for working in any laboratory is never to work alone. You must also use the equipment provided to prevent accidents (such as safety glasses and fume hoods). You need to know the location and understand the operation of safety equipment used to reduce damages from accidents (such as fire extinguishers and safety showers). Opportunities to use safety equipment arise frequently because reactions involving even common chemicals can sometimes do the unexpected. Extreme caution is necessary when working with unfamiliar chemicals.

Several general rules provide guidance for handling chemicals safely.

1. Always wear safety glasses (required by law in many states).

2. Most chemicals used in laboratories are poisonous when taken internally; never taste and avoid smelling any reagent or product. Exceptions to this rule will be rare and clear. For example, if you work on perfumes, smelling the product is a necessary test.

3. Before using any unfamiliar chemical, learn something about its hazards. *READ THE LABEL!* The label on the container will first give you the names of the contents. When reading the names of complex organic compounds, be certain that the name, including each letter and number, identifies exactly the compound you want. For example, 2,2-dimethylhex*ane* is very different from 2,2-dimethylhex*ene* although the names differ by only one letter. Second, the label will give the content's grade or purity. Sometimes the safety, as well as the success of an operation, can depend upon the purity of the reagents. Third, the label will usually state the safety hazards of the material. A typical label is shown in Figure 1.

Never use any chemical found in an unlabeled container.

4. If the material has no indication of safety hazards, **treat it as though it is flammable, volatile, and poisonous** until you know otherwise.

5. If the label on the container doesn't give safety information, obtain the information from some reference sources: your supervisor or a handbook such as *The Handbook of Laboratory Safety* (Chemical Rubber Co.), *Dangerous Properties of Industrial Materials* (Van Nostrand Reinhold), or the *Merck Index*.

6. Avoid skin contact with chemicals. Even the safest organic solvent will probably extract the oil from your skin and leave it dry and itchy. Proper use of tools and transfer methods can eliminate the need for skin contact.

7. Do not clutter your working area with chemicals and apparatus. A neat bench makes working easier and also reduces the chances of knocking over bottles or using the wrong ones. Take chemicals from the stockroom or storage area only when they are needed and return them promptly.

8. Let hot materials cool before putting them into the waste disposal receptacles. Waste disposal instructions are given later.

Ventilation

The words "use adequate ventilation" are often seen on labels or in instructions. This wording is nearly useless. The entire laboratory needs adequate ventilation. If work is in progress, what

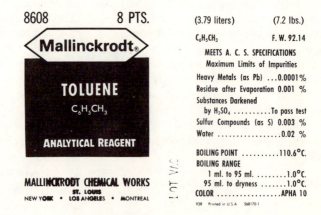

Figure 1. A typical reagent label

is adequate for one operation is completely inadequate for another. If adequate ventilation is recommended for an operation or for the handling of some hazardous material, use a fume hood.

Handling Laboratory Equipment

The most common pieces of simple laboratory bench equipment are illustrated in Figures 2 and 3. Become familiar with their proper names. You will find that your work in a chemical laboratory is greatly simplified if you know the appropriate names and uses of the equipment that is available. This knowledge will help you avoid the misuse of expensive laboratory equipment.

When inserting glass tubing (including thermometers, separatory funnel stems, etc.) into a cork or rubber stopper, wrap the tubing and stopper in several layers of cloth so that both hands will be protected in case the tube breaks. Lubricate the tube and stopper with water or glycerol and use a slight twisting gentle push. Excess force will break the glass, and the jagged ends can make serious cuts. Use similar protection and lubrication when removing tubing from stoppers. Fusiform stoppers help avoid accidents.

Glassware is fragile and requires careful handling. Broken flasks, beakers, cylinders, and test tubes can cause bad spills, ruined experiments, and personal injury. Do not use glassware that is cracked, chipped, or has jagged edges.

Most glassware can be completely cleaned with detergent, water, and a brush. Sometimes a rinse with dilute nitric acid or an organic solvent (which must also be rinsed from the glassware) will help remove persistent films. Do not use concentrated cleaning solutions (concentrated sulfuric or dichromic

acids) unless instructed to do so by your instructor or supervisor. After cleaning and rinsing, glassware may be dried by allowing it to drain and air dry, by warming in an oven, or by aspirating air through the glassware. Heating thick-walled glassware (graduated cylinders, bottles, etc.) in a hot oven or over a Bunsen burner will usually cause them to break.

Pipets, suction filter flasks, and vacuum distillation equipment require special precautions which should be carefully reviewed with your instructor or supervisor when you first use them.

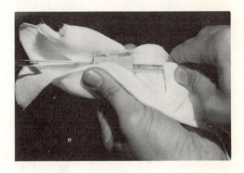

Inserting a thermometer properly

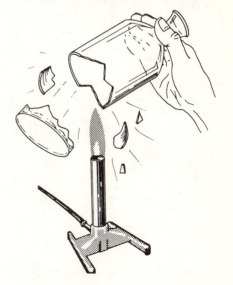

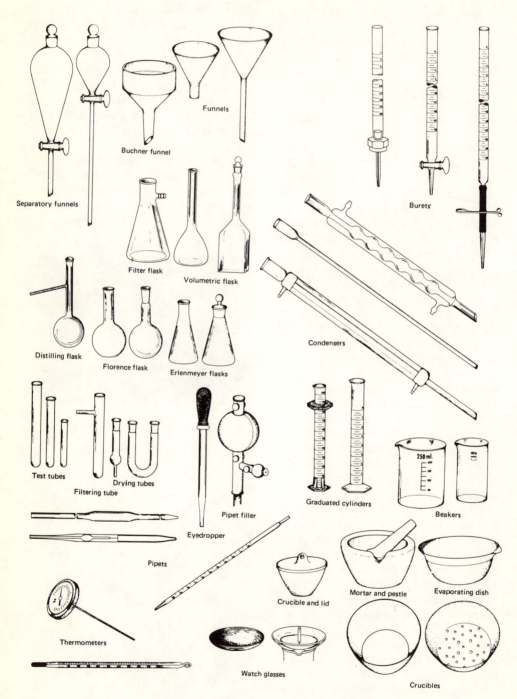

Figure 2.   Chemical laboratory apparatus—glass, polyethylene, and porcelain

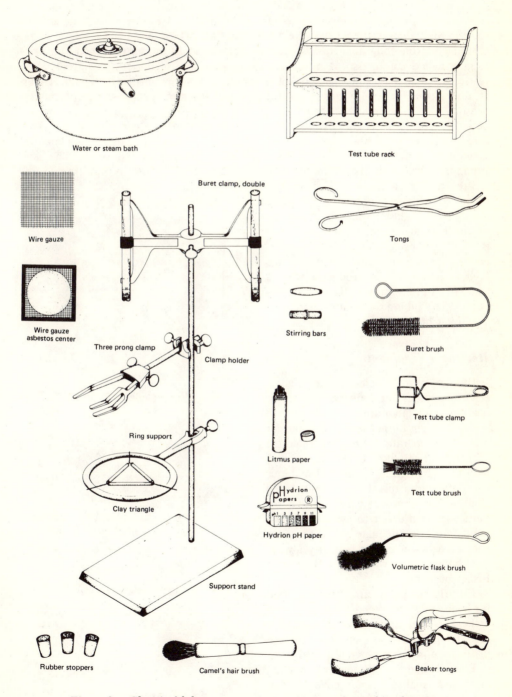

Figure 3. Chemical laboratory apparatus—accessories and implements

There are many kinds of glass used in laboratories. Bottles, ordinary tubing, funnels, and thick-walled glassware are often made from soda-lime glass which is usually called soft glass. It has a low resistance to heat and thermal shock and is not made for operations requiring heating. Soft glass tubing is cheaper and easier to work with at the low temperature produced by a Bunsen burner (bending, drawing, polishing, etc.). Equipment that must be more resistant to breakage and thermal shock is usually made from hard (borosilicate) glass. Although this glass is much more resistant to thermal and mechanical shock, it is more expensive and can be softened enough for bending only at temperatures higher than can be produced with a standard Bunsen burner. The manipulation of glass tubing is explained and illustrated in Chapter 14.

Heating Equipment

Bunsen burners, electric hot plates, electric mantles, or baths are used for heating in the laboratory.

The Bunsen burner and Fisher burner and their operation are illustrated in Figure 4. You should become familiar with them by lighting them and manipulating the adjustments. A small sized Bunsen-type burner called a micro-burner is often easier to use when a low temperature is adequate. A wire gauze with an asbestos center is nearly always used between glassware and a Bunsen or Fisher burner.

Electric hot plates are generally good for heating flat-bottomed containers such as beakers and Erlenmeyer flasks unless high temperatures are needed. Bunsen or Fisher burners are then required.

Electric mantles are electric heating elements, usually asbestos covered, that are molded to be form-fitting around round bottom flasks. The heating rate is controlled by a separate, adjustable rheostat. The mantle must be the correct size to fit properly. When you first use one, get full instructions from your supervisor or instructor concerning proper use and the correct size of mantle and rheostat.

Heating baths can be water (or steam), oil, or sand, depending on the level of temperature desired. The advantage of the baths is the absence of

Hot plate with stirrer

Mantle

Hot plate

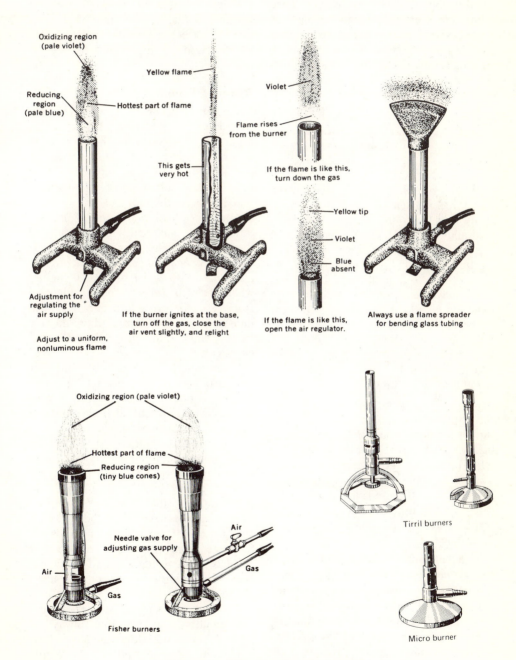

Figure 4.   The operation of a burner

flame or hot electric wires that could ignite flammable vapors. Many electrical heating devices have self-contained adjustable thermostats which spark as they turn off or on. The spark may ignite the vapors from volatile substances causing serious fires. Avoid heating volatile solutions with hot plates, etc., in systems where the vapors can escape.

When a liquid is heated to its boiling point in a glass container, it may not boil smoothly but often superheats in spots and bumps irregularly. In a beaker or Erlenmeyer flask, bumping can be avoided by letting a glass stirring rod touch the bottom of the container directly over the source of heat. However, the addition of boiling chips (small pieces of broken, unglazed clay plate, Carborundum chips, Teflon chips, or glass beads) to a liquid to be boiled is a more convenient practice. Boiling chips must not be added to a liquid at or near its boiling point, or a sudden bump or boil-over may occur. If a liquid has been heated and boiling chips must be added, the liquid must be cooled before the chips are put into the container.

All heating operations must be carefully performed since most organic liquids are flammable. In the presence of highly volatile flammable liquids like ether or alcohol, great care is necessary to prevent vapors from escaping at low temperatures. If there is any doubt about safety, ask your supervisor. In all cases, try to perform the operation under a fume hood.

If a container of solvent does catch fire, *do not* panic. Cover it quickly with a watch glass, evaporating dish, asbestos pad, or something else which will exclude air and snuff out the flame. Do it quickly, deliberately, and carefully. Do not burn yourself in the process! Do not try to carry the lighted beaker or flask. That will fan the flames and make the fire larger. Do not use a fire extinguisher on a simple fire that can be snuffed easily; you may actually spread the fire further. For fires which cannot be snuffed, see a later section for the proper use of extinguishers.

Reagent Containers

An analysis of the impurities was shown on the container label shown in Figure 1. Never do anything that will lower the quality of the material in the container. Therefore, never put a stirring rod, pipet, or other transfer tool into a bottle because the tool may introduce an impurity. Instead, pour some of the material into a small beaker or cylinder and then remove what you need. *Never pour the excess chemical back into the original container.* Close the original container whenever you are not pouring material from it. This procedure prevents dust, moisture, oxygen, and carbon dioxide from changing the properties of many chemicals.

Immediately label the container of any product you have prepared. Be suspicious of sloppy labels. If you cannot be certain of the identification and purity of a substance, discard it.

When reagent bottle caps or stoppers are removed, they can be protected from contamination by holding them in your hand (Figure 5) or letting them lie on clean paper so that only the top and an edge touch the paper. Even then it is best to lay it on a piece of clean filter paper. Stoppers from bottles of corrosive or strongly reactive liquids need particular care in handling to avoid contamination of the stopper.

Concentrated sulfuric acid needs special attention. The acid must always be poured into water slowly and carefully.

Handle reagent bottles of volatile solvents and perform the transfer opera-

A. Read the label twice.

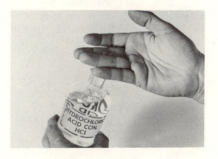

B. How to remove a stopper.

C. NEVER set a stopper down.

D. Pour down a glass rod when possible.

E. When pouring from a beaker, the stirring rod can be held in this manner.

F. Scoop out a little of the material with the spatula provided.

G. Roll and tilt the jar until enough of the material falls out.

Figure 5. Proper methods of removing reagents from appropriate containers

tions so that vapors cannot be ignited or inhaled. Keep solvents away from flames and heat. Solvents and corrosive liquids should be transferred inside fume hoods.

Storage and Transportation

Accidents can be avoided by systematic and orderly storage practices. Tools, reagents, and equipment should be promptly returned to their proper places on labeled shelves and racks. When returning materials to their designated storage area or shelf, double-check the label. Materials that are seldom used clutter benches and floors and are an unnecessary hazard. A proper step ladder or non-slip, non-tip platform should always be used to reach upper shelves or racks. Large or heavy containers or materials that could cause a fire or personal injury hazard if dropped are better stored on lower shelves.

Very large glass containers need extra care when they are being moved. They should be securely grasped by both hands, one under the base and one on the neck. Your hands and the glass should be clean and dry to insure a firm non-slip grip. The containers must be set down very carefully. The best practice to use with large glass containers is to handle them in stands and bottle carriers. When moving glassware or glass containers, always look at the container and in the direction of movement.

Waste Disposal

Safety and health require prompt and proper disposal of waste materials. Carefully consider the *what* and *how* of the material being discarded. Discarding many different kinds of chemicals in the same waste container in the laboratory can result in unplanned reactions, generating foul or highly toxic gases, violent

Always pour H_2SO_4 into water very slowly and stir constantly. Never pour water into H_2SO_4 because enough heat is generated to cause spattering of the hot acid.

explosions, or a boil-over mess. These containers should be only loosely capped (to avoid pressure buildup) and should be stored in a fume hood.

Let hot materials cool before disposing of them.

Do not pour liquids or solutions which do not mix with water into the sink. Do not put solid wastes in the sink.

Flammable liquid wastes and solutions should be collected in designated, labeled, metal containers (preferably U.L. approved) or glass bottles protected by a metal container. This container should not be used for corrosive, reactive, or seriously toxic materials or for highly volatile solvents such as ether, acetone, or carbon disulfide. If special containers are not provided for these kinds of wastes, ask your supervisor about disposal procedures.

Small amounts of water-soluble solvents, not designated as special cases above, may be washed down the sink with large amounts of flowing water; *i.e.,* alcohol wastes. Some individual should be designated to empty the labeled waste containers.

Figure 6.　Solvent waste can

Corrosive materials in small amounts should be neutralized, if possible, before disposal. Acids and alkalies in small amounts may be flushed down the sink with large amounts of flowing water unless it violates state water pollution laws. Larger amounts of acids, alkalies, and corrosive or reactive liquids should not be poured into the general waste disposal containers. They should be put into clearly labeled special containers for proper disposal.

Small amounts of water soluble and nonreactive solid wastes may be dissolved in water and flushed down the sink with flowing water. Otherwise, they should be discarded in the designated crocks or labeled containers. Do not put wet slurries or hygroscopic materials in the open crocks except where such containers are designated for this purpose.

Toxic materials require special disposal methods. Often they are best handled by detoxification through chemical reactions. Then their disposal follows one of the preceding routines. Otherwise, they should be collected in clearly labeled special containers and given to the disposal experts.

With so many variations in materials and means available for disposition of wastes, it is absolutely necessary that a technician think carefully about disposing each chemical waste. Waste scraps of sodium metal, for instance, are solid, reactive, toxic, and dangerous. The reaction products of sodium with air or water are corrosive and flammable. Apparatus which has been used for strong corrosives should be rinsed before being returned to a sink for washing. *Highly volatile, flammable solvents are not safely stored in an ordinary refrigerator which operates through electrical on/off switches that can spark.*

Solvent-wet rags must be discarded in metal containers with tightly-fitted covers. Broken glassware must be disposed of properly so that it will be impossible for anyone to get cut with it.

Machinery

Motors, compressors, vacuum pumps, belt driven machinery, and similar equipment should be protected by sturdy, metal guards. Such protection for per-

Figure 7.   Belt guard

sonnel is frequently overlooked in a laboratory. Exposed belt and pulley systems constitute a serious hazard, and belt guards are strongly recommended.

Two common types of mechanical vacuum pumps are the Cenco Hyvac and the Welch Duoseal. Each pump works by trapping some of the gas from the system to be evacuated, compressing it and forcing out the compressed gas, as shown in Figure 8.

Changing Oil in a Vacuum Pump

The seal between the vane, the rotor, and the frame is maintained by a thin film of oil. This oil must be kept clean if the pump is to work properly. The oil should be changed every three or four months depending on its use. If the oil is dirty or is contaminated with excessive solvents, it should be drained. Then a small amount of new oil should be added, the pump run for a few minutes, and the oil drained. The pump should then be refilled to the mark. Most pumps have a small window on one side to serve as an oil-level gage. Do not overfill the

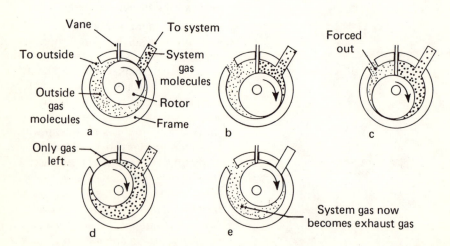

Figure 8.   One cycle of a Hyvac pump

pump with oil. The extra oil will just splash out and make a mess.

CAUTION: When you shut off a vacuum pump, open the system to atmospheric pressure, or the oil will flow back into the system.

Water Aspirators

The need for working with systems at reduced pressure (in the 10–20 mm range) is common. This reduced pressure can be produced with an inexpensive water aspirator attached to a water faucet. The water flow rate is increased until the vacuum is established, and then the water flow rate is reduced until a slow, steady flow of filtrate is obtained through the filtering medium. It is important to check to see that the aspirator has a working check valve in the side arm (Figure 9b). The valve prevents water from being drawn into the system when the water is turned off. Because there is considerable splashing associated with the use of the aspirator, an aerator (Figure 9a) is attached to the outlet. This mixes air with the fast-moving water to reduce splashing. Aspirators operate within a water pressure range of 10–50 psi. When using a water aspirator it is good practice to determine beforehand that a continuous, dependable supply of water at a uniform pressure is available.

Constant Temperature Circulating Baths

In the use of instruments such as refractometers or viscometers where rigid temperature control is required, the constant-temperature circulating bath is used. Thermostatically controlled circulating baths may be either refrigerated or heated. The system may consist of a container to hold the constant-tempera-

ture liquid, a circulating pump or stirrer, thermoregulator with relay box, and a thermometer.

The thermoregulator provides automatic control of the bath at reduced or elevated temperatures. The traditional device consists of a set of glass-coated

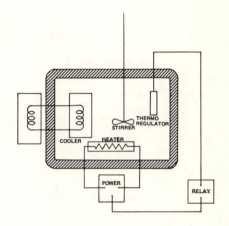

Principal components of a
temperature-controlled bath

b) Water Aspirator

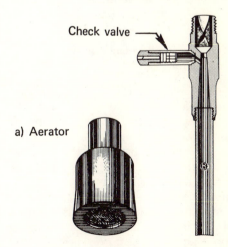

Figure 9. Aspirator check valve and aerator

contact electrodes immersed in separate mercury capillaries in a glass tube (Figure 10). The contacts are sealed to prevent contamination of the mercury surfaces or the electrodes. The setting procedure for a thermoregulator is simple. The sensitivity is 0.03°C, and the temperature range for the devices is from −30°C to 370°C.

The contact electrodes in a thermoregulator have a very low current-carrying capacity. Thus, the thermoregulator is used in conjunction with a power relay. The sensitive coil of the relay is energized or de-energized by a small current flowing through the contact electrodes in the thermoregulator and the relay coil. The relay contacts can carry a current load of 5 amps or more and serve to switch on heaters or cooling units safely.

Solid state devices have greatly improved the accuracy and ease of adjustment in constant temperature baths; such devices may be accurate to 0.001°C. The solid state thermostat consists of four major parts—a sensing element, a control amplifier, the heating or cooling element, and the controlled component, *e.g.,* an air or liquid bath.

Temperature Sensors and Controllers

Solid state sensors may include thermocouples, thermistors, or bimetallic elements. A thermocouple is a comparison device which responds to temperature differences between two junctions of dissimilar metals. Thermistors are temperature-sensitive resistors. A bimetallic element bends as a result of temperature change and opens or closes electrical contacts. In baths designed for temperature control, heaters may consist of lightbulbs, infrared lamps, or electrical heating coils. Cooling baths

use refrigeration coils similar to those in a home refrigerator. Insulation may be required to make the bath operate efficiently.

The proper positioning of the thermoregulator sensor, heater, and stirrer within the bath is critical. The three components should be relatively close to each other to make space available

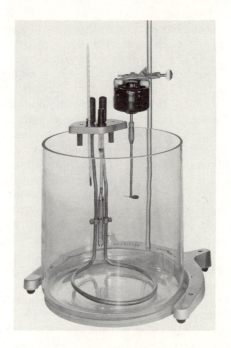

Figure 10.   Thermoregulator in a temperature-controlled bath

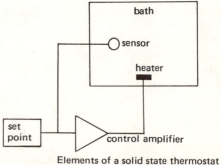

Elements of a solid state thermostat

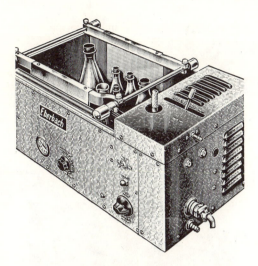

Temperature-controlled water bath shaker

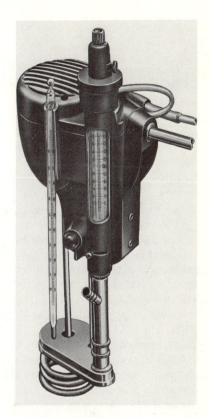

Figure 11.   Self-contained, constant temperature circulator

within the bath for the system to be studied. The sensor must be guarded from the heater so it will respond to the overall temperature of the bath rather than to changes near the heater.

Stirrers driven by a motor are used in most temperature controlled devices. One brand of bath uses a powerful magnet underneath the unit to move an agitator plate in the bath, causing a gentle stirring throughout the bath. Some baths contain a shaking mechanism with special holders for flasks, test tubes, etc.

Water is generally the bath liquid for temperatures between 0 and 60°C. Above that temperature, silicone oil, petroleum oil, or vegetable oil is used.

Self-contained, constant temperature circulators are available. These units can convert any suitable container to a constant temperature bath or circulating system. The device has a pump for circulating a controlled liquid through the bath or to an external instrument, *e.g.,* a refractometer. The electric heating element, thermoregulator, relay control, and thermometer are housed in a compact case which can attach to a ringstand or to the edge of the bath. Commercial units control baths in the range from ambient temperature to 100°C with 0.01°C accuracy.

Do Not Work Alone in a Laboratory!

3

Personal Protective Equipment

When working in the laboratory, garage, or home workshop, we should try to avoid accidents. Laboratories are safer than most homes even though potential hazards are greater. The reason for the safety is that everyone who builds, equips, manages, or works in a laboratory recognizes the potentially dangerous situations and takes the necessary precautions to prevent accidents. Protective equipment must be used frequently to prevent accidents or to reduce their seriousness when they do occur.

All accidents should be reported to your supervisor. A regular channel for reporting accidents is required by some insurance companies. A complete, accurate report helps to determine the facts so that similar accidents can be prevented in the future.

Laboratory Safety Equipment

Your eyes are irreplaceable; and they can be injured easily by corrosive or hot chemicals and flying pieces of exploded apparatus. Always wear safety glasses whenever there is any possibility of eye injury. When chemical work is being done, many places will require that safety glasses be worn. If you normally wear glasses, it is strongly advised that you get and use prescription safety glasses. If you don't normally wear glasses, get into the habit of wearing safety glasses when-

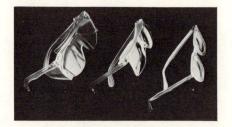

Safety glasses

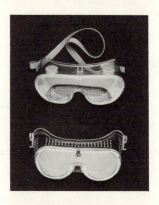

Safety goggles

Face shield

ever you are in a laboratory area, even if you are not doing experimental work.

There are many different designs of safety glasses, and they provide different amounts of eye protection. Safety glasses are usually made of transparent plastic or of hardened safety glass. Some have head bands or hooks for your ears to keep them in place if they are hit. Some have side shields to protect against chemical splashes while others have full shielding all around. Some goggles are made to fit over regular glasses; some are made to protect your eyes from ultraviolet radiation. Choose the best eye protection for the work you are doing and then wear it. Remember, however, to shield any source of trouble. Safety glasses are intended to be the last line of defense.

A safety shield permits observation of an apparatus while providing protection for all of your body. It also protects the laboratory from exploding glassware. The portable shield of shatterproof glass, safety plate glass, or strong transparent plastic is simply placed on the bench in front of the potentially dangerous apparatus. When standing in front of the shield, you are protected from explosions. However, you can still see the apparatus and manipulate it by reaching around the sides of the shield.

Since laboratories regularly use poisonous, corrosive, and flammable materials, special safety equipment is placed in or near each working area. Proper use of this equipment will usually reduce the damage or injury caused by an accident. When an accident occurs, it is most important that the right equipment or treatment be applied as *quickly as possible*. Always make sure you know where the special safety equipment is kept, when it should be used, and how to use it.

Safety Showers and Fountains

The most important pieces of special apparatus for your personal protection against injury from fire or corrosive chemicals are the safety shower and the eye wash fountain. (Some laboratories use eye wash hoses rather than fountains.) If your clothing ever catches fire or if a corrosive chemical is spilled on you, *every second will count* in saving yourself from serious injury. If such a moment ever comes, you may be confused by pain or panic. Always note the exact location of the safety shower and eye wash fountain when you start work in any laboratory area. You must be totally familiar with the procedures needed to operate them.

Safety showers are usually operated by pulling on a chain or ring. If you have clothing that is burning or wet with a corrosive liquid, get under the shower head and pull the ring. Water will flow at a very high rate (much stronger than a home shower) and continue to flow after you release the ring. If eyes are affected, look up toward the shower head to flush them. If your clothing is soaked with a toxic or corrosive material, remove your clothes. Lost modesty is less painful than lost skin!

If you get any corrosive or irritating chemical in the eyes or on the face, go quickly to the eye wash fountain, remove

Safety shield

Safety shower and eye wash fountain

your safety glasses, and bend over until your eyes are directly in the dual streams of water. Rinse thoroughly until all of the foreign material is removed. Fifteen minutes is the recommended minimum time.

Fire Control

Fire extinguishers inside laboratories are usually mounted on walls near entry doors. A better location is in the hall just outside the entry door. This puts them at the most convenient location for access during an emergency retreat or entry in case of sudden need. You should know what kinds of extinguishers are available and be familiar with their operating instructions.

The most common extinguisher is filled with carbon dioxide. It may be used on both electrical and chemical fires. The operation of the particular extinguisher will be described on the label. Operation of the carbon dioxide extinguisher involves just 4 steps:

1. Remove the unit from its mount and bring it close to the fire area.
2. Pull the handle retainer ring.
3. Point horn at the base of the fire.
4. Squeeze the handle.

A blast of CO_2 gas and solid will snuff out the flames. A carbon dioxide blast may spread the fire if a pool of burning liquid is involved. Such fires can best be fought with a dry chemical or foam type extinguisher which is designed to be used on gasoline, oil, or similar fires. A dry chemical extinguisher operates much like an aerosol spray can. Burning liquids are more effectively extinguished by the ejected stream of dry sodium bicarbonate powder than by carbon dioxide blasts.

Water pumps, soda-acid extinguishers, and fire hoses are often available for fighting fires of paper, wood, or other common combustibles. They are *not* good choices for some chemical fires or for any electrical fires. Water is the best choice for extinguishing burning clothing.

Sand-filled buckets are often kept in laboratories. They are useful for smothering small fires.

Laboratories that use chemicals such as potassium or lithium will be equipped with special fire control devices, and

Fire extinguisher

laboratory workers will be specially trained in their use.

Other Safety Equipment

Acid and alkali burns on the skin require *immediate* washing with large amounts of cold water. Most laboratories also have saturated boric acid solutions to neutralize alkali burns. They will have sodium bicarbonate (baking soda) solutions to neutralize acid burns. These solutions should be loosely covered to prevent contamination. They are not substitutes for very thorough washing with water, but they are helpful in neutralizing the remaining acid or alkali that may continue to stick to the skin.

Most laboratories will be equipped with a *general first aid kit* to provide quick treatment for minor burns and cuts. In addition, the kit may contain medications to neutralize the effects of some toxic materials. They may also contain chemicals to induce vomiting when desirable. Most kits will contain substances to help revive fainting victims. The containers will be clearly marked and provide instructions for how and when to use them.

Fume Hoods

Virtually all chemical laboratory facilities will contain fume hoods. Hoods are designed to exhaust toxic, flammable, and unpleasant vapors and dusts rapidly. Before starting an experiment, determine if the chemicals you will use require that the work be done in the fume hood.

The hood will be similar to most standard laboratory benches. It will usually include a sink, water spigots, and gas and electrical services. The major difference between the standard laboratory bench and the hood is that the hood area is enclosed on three sides by a solid barrier and at the front by a safety glass shield which may be raised or lowered.

These will shield against minor explosions but not against detonations. Special hoods are made for special hazards.

When you get ready to use a hood, check to see that it is not being used to store hazardous chemicals. Hoods are frequently used for storage, but this is a questionable safety practice. Be sure the air exhaust system is operating properly. To check the exhaust system, turn on the fan and close the front shield to within one inch of the bench top. The air flow into the hood should now be strong enough to cause paper to "flap in the breeze" when it is held in the opening. It is also important that good ventilation into the hood is maintained with the front shield opened as it might be while an operation is being performed.

Arrange the apparatus in the hood in a way that permits you to watch the apparatus and operate controls, stopcocks, funnels, etc., without reaching over or around flames or heat sources. In an emergency you must be able to turn off gas and electrical devices without leaning into the hood.

When your equipment is in operation and you no longer need to manipulate it, lower the sliding front sash to within a few inches of the bench. Check your equipment to be sure that the increased air velocity resulting from lowering the sash doesn't blow out your burner flame or move lightweight objects.

The fact that you are using a hood does not mean you can neglect other safety procedures and equipment. If the work in the hood involves highly toxic, flammable, or explosive material, safety shields should be used. Keeping a fire extinguisher and gas mask nearby and ready for use is a good safety practice. When you are finished using the hood, dismantle the equipment and return it to proper storage areas. Do not use the hood as a chemical storeroom.

Fume hood

Volatile materials that are radioactive must not be used in ordinary fume hoods. Their handling will be discussed in another section.

Exhaust Fans

If exhaust fans are used to ventilate your work area, make sure they are running when you work in the laboratory and that they are not working in opposition to the desired hood flow. *They are not substitutes for hoods.* Frequently, exhaust fans will not remove all toxic or flammable vapors from the work area.

Breathing Apparatus

In addition to receiving burns, many people are overcome by smoke or toxic fumes during fires. Special gas masks should be located in or near the laboratory for use during fire fighting and for entering spaces containing toxic or irritating vapors. The masks are usually stored in special red containers with black lettering. When needed, the mask should be taken from the container and checked to make sure it is suitable for protection against the particular toxic vapors present. If it is a suitable mask, remove the seal from the bottom of the canister (the rectangular or cylindrical can attached to the mask) and fasten the mask over your face with the straps provided. Inhale and check the tightness of the mask by placing your hand over the inlet of the canister. (This is the opening from which you removed the cap seal.) When you do this, the mask will either press tightly against your face or you will feel air seeping through a leak around the mask edge. If it leaks, tighten the straps and test it again. Note: A tight fit of masks over beards or bushy sideburns is difficult and sometimes nearly impossible.

Though the canister mask is the most common and is very useful, there are other types of breathing apparatus. Filter respirators are frequently used by painters and workers who handle powders and dust-forming materials. These devices are not gas masks and have no value in toxic vapor, smoke-laden atmospheres, or in atmospheres deficient in oxygen. They simply filter dust and mists out of air that is being inhaled.

A self-contained breathing apparatus has its own air or oxygen supply. This equipment is used when the oxygen content of the air has become too low. A standard canister mask can only remove toxic vapors; it cannot supply oxygen if none is available. Normally, only specially trained people will use self-contained breathing apparatus. A newer type of mask in which oxygen is generated from a potassium oxide canister is becoming common. To use this mask remove the paper canister cover from the bottom, place the mask over your head, tighten the straps, and breathe normally. A fresh cartridge must be inserted after every use.

Canister mask

Laboratory Clothing

Laboratory coats and aprons should be used for safety and convenience. They will prevent spilled or spattered chemicals from injuring you or destroying your clothing. Lab coats also contain pockets for holding small items used in the laboratory. Special safety clothing is provided when needed for protection against biological, radiological, or special chemical contamination.

Some laboratory operations require handling or being near gases, liquids, or solids which are skin irritants, allergy producers, or cancer producers. Elastic gloves are excellent for protecting your hands from such materials. In special cases protective skin creams may be useful in preventing skin irritations. When hot objects must be handled or closely approached, asbestos gloves are used. Choose proper gloves to minimize the hazard. Rubber or plastic may dissolve and are worthless with very hot objects. Asbestos can soak up spilled chemicals and increase the hazard from a chemical spilled on one's hand.

When technicians work in pilot plant or production plant locations, a hard hat helmet and hard toe safety shoes are usually required. They can effectively protect your head and feet from falling objects.

Visitors

Visitors or guests in a laboratory should be provided with any necessary protective equipment.

4

Fire Safety

As a chemical technician, you should know how to prevent fires and how to extinguish them when they do get started. Even more than this, you need to know about the science of fire. Why do some things burn while others do not? Why do some liquids and solids catch fire easily while others are difficult to start burning? Why do some gases seem to explode when ignited? You are much safer in a laboratory when you know the answer to these questions.

Flammable Liquids and Gases

Things "burn" only when exposed to a source of oxygen. (There are a few situations in which burning occurs without oxygen. In this book, our discussion of fire will assume that oxygen makes burning possible.) The oxygen can come from solids, liquids, or pure oxygen as well as from the air. Removing the oxygen from air leaves nitrogen and other inert gases that can be used to extinguish or smother fires.

Burning (combustion or oxidation) is a reaction between the burning substance and oxygen. *Combustion* is defined more exactly as the rapid oxidation of a substance which produces both light and heat. This definition enables us to distinguish between fire conditions and slow oxidation such as the rusting of iron and

life-supporting reactions inside your body.

Substances that burn are subdivided for technical reasons. We need to direct our attention to those subdivisions that are important to chemical technicians.

CAUTION: Never experiment with flammable liquids and gases of explosive solids or mixtures except under the direct and careful supervision of someone who knows exactly what is being done, what hazards are involved, and how to avoid accidents.

Only gases burn. Liquids that appear to be burning are really vaporizing; the vapor mixes with air and burns. Before burning, a solid must be converted to a gas. It is actually the gas or vapor that burns.

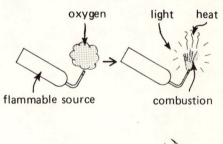

flammable source → combustion

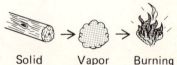

Solid → Vapor → Burning

Processes or reactions that are properly called *explosions* or *detonations* should not be confused with combustion. Explosions occur when gases are already mixed with air or oxygen and when liquids or solids have an oxygen composition sufficient to oxidize the rest of the substance. Since the oxygen is thoroughly mixed with the substance to be oxidized, the reaction is rapid and violent. These reactions usually produce a very large volume of hot gases. Burning or ordinary combustion takes place at a much slower rate and is more easily controlled.

Flammability Limits

Gases that will burn have flammable or explosive limits. A gas, like gasoline vapor, will burn only if mixed with oxygen or air. Not all mixtures of air and gasoline vapor will ignite. The test for establishing the flammable or explosive limit determines if the mixture will be ignited when exposed to a flame, hot wire, or spark. If the mixture catches fire and the flame spreads through its entire volume (called propagation of flame), the mixture is within its flammable limits. If there is not enough gasoline vapor in the air to burn, the mixture is said to be too lean. If there is too much vapor to air to allow the mixture to ignite and the flame to propagate, it is said to be too rich to burn. There must be at least 1.3% gasoline vapor to 98.7% air at the lower limit, and there can not be more than about 7% gasoline vapor to 93% air at the upper limit or the mixture will not ignite and propagate flame. A pure gasoline vapor or a mixture of gasoline vapor and air that is too rich to burn (more than 7% vapor) will usually mix with enough additional air to get within the flammable range limits and can be ignited. This is

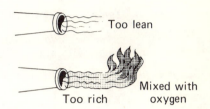

why the vapor over an open container of gasoline seems to burn somewhat above the surface and why a jet of gasoline vapor burns a short distance away from the nozzle.

Gasoline vapor, although not a pure substance, has served as a familiar example, but gasolines can vary in composition and in flammable limits. All pure flammable gases have definite upper and lower flammable or explosive limits. So do the vapors of flammable liquids and dusts from solids that will burn. However, the limits vary widely among substances. Examples are given in Table I.

If a source of ignition, such as a flame, is brought close to the liquid surface of a flammable liquid in a cold, open container, one of three things may occur: 1) nothing may happen, 2) there may be a flash of flame but no continuation of burning, or 3) it may ignite and continue to burn. With some liquids a burning match stick can be extinguished in the liquid without any flash or burning. If a flammable liquid is heated, there will be some temperature range at which there will be a flash but no continued burning if a flame is brought near the liquid's surface. At some higher temperature, the same kind of flame may cause the vapor to catch fire and continue to burn.

There are three definitions to describe these characteristic actions. The **flash point** is the lowest temperature at which a substance or mixture in an open vessel gives off enough combustible vapors to produce a momentary flash of fire when a small flame is passed near its surface.

Table I. Fire Information for Selected Gases (*1*)

Gas or Vapor	Flammable Limits, volume % in air		Flash Point, °C[a]	Auto-Ignition Temperature, °C
	Lower	*Upper*		
n-octane	1.0	6.5	56	428
n-hexane	1.1	7.5	—7	437
benzene	1.3	7.1	12	1040
acetylene	2.5	80—100		581
carbon monoxide	12.5	74		1128
acetone	2.6	12.8	0	869
ethyl ether	1.9	36.0	—49	320
propane	2.2	9.5		842
methane	5.0	15.0		1004
hydrogen	4.0	75		752

[a] For closed cup *vs* open cup procedures see NFPA No. 325 M.

The **ignition temperature** is the lowest temperature at which the vapor over the liquid will ignite and continue to burn if an ignition source (flame, hot wire, spark) is applied near the surface. The **auto-ignition temperature** is the lowest temperature at which a vapor mixed with air will ignite spontaneously without any outside source of ignition.

Tabulated flammability limits are usually given for room temperature and standard atmospheric pressure. Different experimental conditions and test methods will give different results. Also some literature references do not clearly state whether ignition temperatures or auto-ignition temperatures are used. Flammable limits vary with temperature and the concentration of oxygen (*see* Table II).

Flammable and explosive limits should be used as a guide for whether a high, medium, or low hazard exists. It is best that you assume that once the flash point is reached, the liquid may catch fire and continue to burn. For example, ethyl ether is a very serious fire hazard. Any experienced chemical technician knows that an open container of ethyl ether will ignite very readily at temperatures well below the auto-ignition temperature of 320°C. If the vapors of flammable liquids are allowed to escape around desk tops or into sinks, they may catch fire, explode, or *flash back* (flame goes back to the source of the vapor) and cause a major fire.

A liquid in an open container burns with a flame much hotter than its ignition temperature. Thus, the surface is heated, vaporization is faster, and burning is accelerated.

Most flammable vapors are heavier than air and will flow along desk tops, floors, or troughs. Dangerous vapor concentrations can collect in low spots, such as sinks and sumps, and produce an invisible hazard. Hydrogen and natural gas are exceptions and tend to rise rather than sink. Empty containers may still be full of vapor and must be properly vented or handled very carefully.

National Fire Protection Association

The National Fire Protection Association is an organization devoted to the study of fire accidents, fire prevention, and safety. It proposes rules and ordinances for use by industrial and public

Table II. Influence of Oxygen Concentration on Flammability

| | *Flammable Limits, Volume %* | | | |
| Substance | *In Air* | | *In Oxygen* | |
	Lower Limit	*Upper Limit*	*Lower Limit*	*Upper Limit*
Carbon monoxide	12.5	74.2	15.5	93.9
Ethyl ether	1.7	48	2.1	82.0
Ammonia	15.5	27.0	13.5	79.0
Hydrogen	4.1	74.2	4.6	94

authorities. The NFPA annually revises and publishes the National Fire Codes. Individual pocket edition pamphlets (NFPA Standards) covering most of the subsections of the National Fire Codes are available. These cover a wide variety of subjects about recommended practices in construction, equipment, and other aspects of fire prevention and protection. One of the most useful pamphlets for a chemical technician's personal library is NFPA Standard 325 M, *Fire Hazard Properties of Flammable Liquids, Gases, Volatile Solids*. The NFPA publishes many other periodicals and reports which should be known to anyone actively connected with fire prevention and protection.

Liquid Flammability Classifications

The NFPA has prepared the following widely accepted definitions: **combustible liquids** shall mean any liquid having a flash point at or above 140°F (60°C) and below 200°F (93.4°C) and shall be known as Class III liquids. Note: The upper limit of 200°F (93.4°C) should not be construed as indicating that liquids with higher flash points are non-combustible. **Flammable liquids** shall mean any liquid having a flash point below 140°F (60°C) and having a vapor pressure not exceeding 40 psia (pounds per square inch, absolute) at 100°F (37.8°C). Flammable liquids shall be divided into two classes of liquids as follows:

Class I liquids shall include those having flash points below 100°F (37.8°C) and may be subdivided as follows:

Class IA liquids shall include those having flash points below 73°F (22.8°C) and having a boiling point below 100°F (37.8°C).

Class IB liquids shall include those having flash points below 73°F (22.8°C) and having a boiling point above or at 100°F (37.8°C).

Class IC liquids shall include those having flash points at or above 73°F (22.8°C) and below 100°F (37.8°C).

Class II liquids shall include those having flash points at or above 100°F (37.8°C) and below 140°F (60°C).

Note: The volatility of liquids is increased when they are heated to temperatures equal to or higher than their flash points. When so heated, Class II and Class III liquids shall be subject to the applicable requirements for Class I or Class II liquids. This code may also be applied to heated high flash point liquids even though they are outside its scope when not heated.

Unstable (reactive) liquid shall mean a liquid which, in the pure state or as commercially produced or transported, will vigorously polymerize, decompose, condense, or become self-reactive under conditions of shock, pressure, or temperature.

Portable vapor detecting meters, or sniffers, are available commercially and may be calibrated to indicate the relationship between an atmosphere containing vapors and the flammable limits for that vapor.

Storage of Flammable Liquids

In laboratories, flammable liquids are usually stored in glass bottles for convenience and to prevent contamination. However, bottles break and cause serious hazards. Flammable liquids should be stored in the smallest bottles consistent with the usage, usually not more than one liter of liquid per bottle. When practical, metal containers are preferred from the standpoint of safety, however, they present hazards from corrosion. Special safety cans with spring-loaded spout covers are excellent for non-corrosive liquids. When flammable liquid containers must be larger than one liter, there are special rules for storage and handling. Usually these containers are put in rooms with special ventilation and construction. Some storerooms and laboratories insist that all glass bottles of flammable or corrosive liquids larger than one liter be protected by metal or plastic catch pans or containers.

Fire Prevention

Great care must always be used to prevent the escape of flammable gases and liquids to the environment. All containers, whether open or closed, should be protected from heat. All sources of ignition must be kept completely away from flammable substances. Sources of ignition include open flames, hot wires, electric sparks from switches or motors,

steam baths, and hot steam lines (which can ignite carbon disulfide). Static electricity sparks and some oxidizing agents are unexpected sources of ignition under certain conditions. Any fire needs fuel, oxygen, and a source of ignition. Constant attention toward preventing these three elements from coming together will help eliminate fire accidents.

Extinguishing fires has been thoroughly discussed in the last chapter. Chemical technicians should be familiar with the proper use and applicability of water, dry chemical extinguishers, carbon dioxide extinguishers, and fixed systems that discharge water, carbon dioxide, or nitrogen into high hazard areas.

Gas fires can often be extinguished most effectively by eliminating the source of the gas rather than trying to stop the fire. Even with the flame extinguished, a continuing flow of gas will regenerate a fire hazard, or worse, an explosion hazard.

Container labels describe fire hazards —read them!

Pyrophoric Materials

Finely divided dust suspended in air (like flour or coal dust or finely ground metal or paint pigments) can act as if it were gas. In the correct proportions with air and with a source of ignition, dust suspensions can explode violently.

Technicians in laboratories and plant operations should be aware of pyrophoric materials. These are materials which will heat and ignite spontaneously in the presence of air. They can unexpectedly become a source of ignition in the presence of flammable gases and vapors. An example is the metal sulfide deposits that may collect on the inside of large steel tanks storing sour (sulfur-containing) refinery oils and refining intermediates. Where pyrophoric materials may occur,

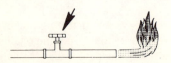

special rules will usually be applied and must be observed. A film, "The Science of Fire," convincingly shows that rag and paper towels wetted with oils, paint, or other flammable liquids should be discarded in air tight metal containers to prevent them from igniting spontaneously upon exposure to air.

A chemical technician must study the potential hazards connected with any flammable gases or liquids that he may use. He must know the characteristics of these substances so that he can handle them intelligently and protect himself and his co-workers from fire and explosion accidents.

Literature Cited

1. National Fire Protection Association, "Fire Hazard Properties of Flammable Liquids, Gases, Volatile Solids," (1969), NFPA No. **325M,** 470 Atlantic Ave., Boston.

5

Toxic Chemicals

Why are there repeated warnings throughout training programs for chemical technicians about the hazards of toxic (poisonous) chemicals? The main reason is that many hazards presented by these chemicals are not evident from appearance, smell, or everyday knowledge. Toxic effects often take a long time to notice and then it may be too late to take remedial measures. Hazards must be foreseen and avoided. You need a thorough knowledge of toxic chemicals and how to use them safely. Always remember that *avoiding accidents depends upon your constant personal attention.*

Hazardous Chemicals

It is safest to assume that all chemicals, even water if not safely handled, can be hazardous. Become acquainted with the possible hazards of every chemical material that you use. The teamwork aspect of most chemical laboratory work has been stressed. Nowhere is teamwork more important than when handling hazardous materials. One careless individual may expose himself and many others to dangerous situations. Carelessness in a chemical laboratory simply cannot be tolerated, and a frequent offender will be dismissed. *Your first careless act could also be your last.*

Hazards associated with flammable gases and vapors, explosive chemicals, and other physically hazardous materials are described in other sections. The dangers discussed in this chapter are associated with toxic chemicals and the physiological effects arising from over-exposure to them. We will also discuss detection and protection against a dangerous environment.

Serious exposures can also occur outside the laboratory, such as in chemical plants and even in your home, garage, or car. Fortunately, the increased opportunities for exposure in the laboratory lead to greater awareness. Thus, more careful attention is given to precautions that make the laboratory and the plant safe places to be. However, this result occurs only if you are constantly alert; neither carelessness nor slip-ups are permitted.

In any industrial environment where dangers exist, there will usually be a safety organization. While directives must be followed, the ultimate responsibility rests on the individual. The user must recognize any and all hazards involved with the particular material he is handling and act accordingly.

Toxicity

Herbs and extracts have been used to treat disease since ancient times. Thousands of years ago, people had some knowledge of the physiological effects of certain materials, such as naturally occurring plant and mineral poisons and probably even mercury. In the 16th century, Paracelsus described the toxic nature of mercury and prescribed a treatment for mercury poisoning.(*1*) In Austria the hazards in mining mercury were recognized as early as 1665.(*2*) Throughout history, appreciable numbers of workers are reported to have been victims of overexposure to chlorine, arsenic, thallium, hydrogen sulfide, selenium, and scores of other toxic elements or compounds.

Many toxic chemicals are produced every year. These include fungicides, insecticides, fumigants, disinfectants, rodenticides, repellants, herbicides, and defoliants produced for the agricultural and livestock markets. All possible precautions must be exercised in the research, development, manufacture, shipping, and use of these chemicals. When the toxicity of chemicals is not thoroughly respected, accidents can and do happen.

The Manufacturing Chemists' Association, has adopted the following definitions that are applicable to toxic chemicals:(*3*)

toxicology—the science which studies the ability of a substance to cause bodily harm by other than mechanical means

toxicity—the capacity of a substance to cause injury or harm to living tissue

hazard—the probability that injury will result from handling or using a substance in the quantity, frequency, and manner proposed

systemic—reference to the organ systems of the body; conditions involving the body as a whole

local—reference to a limited, well-defined area of injury or response. In common usage this may refer to surface effects.

topical—refers specifically to surface application or contact

acute—refers to a single or sudden event or response; thus, an acute exposure or an acute illness

chronic—refers to a repeated or prolonged event or response; thus, a chronic exposure or illness.

threshold limit value (TLV)—This refers to a concentration of dust, mist, or vapor believed to be harmless and unobjectionable to most humans when they are continuously exposed for an eight-hour working day, five days a week. There is usually a sufficient margin of safety to permit exposures in excess of eight hours without harm. Exceeding the threshold limit of exposure does not necessarily mean that employees will be harmed, but it does imply a greater probability that one or more will be harmed or will complain of sensory or subjective effects in the course of time.

Entry of Toxic Chemicals into the Body

The three most common ways in which toxic chemicals enter the body are by inhalation (breathing), ingestion (swallowing), and absorption through the skin. Of the three, the most frequent means is by absorption, but the entry method causing the highest percentage of fatalities is inhalation, as shown by Table I.

Table I. Disabling Work Injuries in California, 1963 (4)

(From contact with radiation, caustics, toxic, or noxious substances)

Contact Method	Total	Fatal	Non-fatal
Inhalation	1203	23	1180
Absorption	4123	3	4120
Ingestion	145	2	143
Other	41	11	30

Respiration and the Metabolic Balance

In order to understand how various toxic materials can interfere with normal body functions, it is helpful to recognize certain delicate but vital forms of equilibrium in the human system.

Oxygen (with a concentration of about 20% in normal air) is inhaled and passes through the respiratory passages, becoming somewhat conditioned and filtered before reaching the bronchial tubes. In the lungs the air passes into smaller and smaller tubes, ending in tiny air sacs. The latter are surrounded by a capillary network that carries the blood supply. In these air sacs the air and the blood are separated by very thin walls. It is here that an equilibrium is established. The blood absorbs oxygen and releases carbon dioxide. In each case the gases pass from an area where they have a high concentration to one where their concentration is lower. The difference in concentration determines the rate of oxygen intake and carbon dioxide release. Any upset to this equilibrium can be a very serious matter.

The oxygen in the blood is transported to other parts of the body where it is released within the cells for normal life processes. As the oxygen is consumed, the principle oxidation products are water and carbon dioxide. The blood, having lost its oxygen and now carrying the carbon dioxide, returns to the lungs to discharge the carbon dioxide and to get a fresh supply of oxygen. For the right balance to be maintained, a sufficient supply of fresh oxygen must be available in the lungs, and the pH of the blood must be right to allow the carbon dioxide to be released.

The balance will be upset by an abnormal condition such as oxygen deficiency. If the blood cannot absorb enough oxygen or cannot release it to the body's cells, the blood's pH changes and the carbon dioxide concentration is affected. If the partial pressure of oxygen falls below 120 torr (in normal air it is 150 torr) the situation becomes critical. The body can adapt to a low oxygen concentration by manufacturing more red blood cells, but this requires a long time and is no help in emergency situations.

Normally all a body's systems cooperate to maintain the essential balances. However, a disease or the introduction of an unexpected chemical influence can produce a disastrous effect. In some cases a signal to the body to increase the breathing rate makes a bad condition worse, and the result may be fatal.

Physiological Reactions to Toxic Chemicals

The extent of a human body's reaction upon exposure to toxic chemicals is determined by many factors. The most important are:

1. The *concentration* of toxic material in the blood system or in a specific, sensitive organ.

2. The body's *tolerance* to a particular toxic element.

3. The *rate* at which the toxic material can be consumed, detoxified, or eliminated before irreparable damage is done.

The level of hazard depends on the concentration of the chemical in the environment, the length of exposure, the general health of the individual exposed, and the success and speed of remedial measures once the exposure is recognized. These considerations must be appropriately interpreted for cumulative poisons (poisons that the body absorbs and builds up over a period of time) since it is rare that a body develops increased tolerance for a toxic chemical as a result of continued low-grade exposures. Factors affecting the severity of the reaction are different for poisoning through inhalation, absorption through the skin, or ingestion.

Many bodily responses to poisonous chemicals are the same although the chemicals may enter in different ways. Some of the responses that may develop are:

1. Alteration of the composition of the blood
2. Internal coagulation of blood
3. A change in the number of red blood cells and amount of hemoglobin
4. A change in the pH or viscosity of the blood
5. Rapid destruction of cells
6. Change in rate of blood circulation
7. Change in breathing rate
8. Change in blood pressure.

Some of these changes may alter the functioning of specialized organs such as the heart, kidneys, liver, brain, or lungs.

The body will attempt to alter foreign substances by one or more of several methods. However, it is not generally well-prepared to do this. As a substitute for its inability to detoxify a poison, the body might try to establish a new equilibrium that will tolerate the intruder until it can be eliminated. The earlier description of the delicacy of equilibrium balances in the body makes it plain that a new equilibrium cannot usually be established without other problems arising. An example is the way the body tries to cope with an excessive amount of ethyl alcohol.

Threshold Limit Values

Boundary limits have been set for toxic materials so that atmospheric concentrations that will cause bodily harm can be prevented. These limits are called *threshold limit values* and have been established for toxic gases, vapors, mists, dusts, and fumes. Atmospheric situations are usually easier to evaluate and control than the circumstances of poisoning by skin contact or swallowing. The latter conditions are usually caused by an accident rather than by continuous exposure. For example, it is known that if 12 drops of a 50% solution of Phosdrin remain on the skin, the poisoning will almost certainly be fatal to an adult.(5) The proper protection is not to make sure that fewer than 12 Phosdrin drops fall on the skin but to prevent any Phosdrin from touching the skin. Many people regularly work and live in atmospheres containing vapors that are known to be severely toxic in high concentrations but which are not hazardous in low concentrations.

The threshold limit value or TLV (sometimes called maximum allowable concentration or MAC) is defined on page 33 and is further explained in the footnote of Table II. Gases and vapors usually have their TLV expressed as parts per million (ppm) parts of air. Toxic dusts, mists, and fumes are usually expressed as milligrams per cubic meter of air although mineral dusts such as silica may have threshold limits expressed as millions of particles per cubic foot of air.

Exposure by Inhalation

Some vapors or gases, like ammonia, give warning of their presence by their odor or immediate irritating effect before the concentration rises beyond the TLV. Others, such as carbon monoxide, have no odor, taste, or other warning symptoms. Still others, like hydrogen sulfide, have an easily detectable warning odor that soon becomes unnoticeable, and a deadly concentration can develop without the exposed person's being aware of it. Asphyxiation can occur without warning if oxygen concentration becomes too low or carbon dioxide concentration becomes too high.

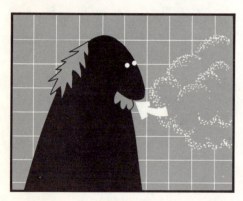

Individuals vary widely in their symptoms and reactions when exposed to vapors of toxic chemicals. The most harmful effect is usually absorption of the vapor by the blood. This results in rapid distribution of the poison throughout the body and leads to the upset of body systems characteristic for the particular toxic material. The functioning of particular organs may be seriously upset. Frequently the first symptoms to appear are blurred vision and malfunc-

tion of muscles. Whatever the first noticeable symptoms are, some disturbances may have already progressed so far that complete recovery is impossible. Kidney failure caused by overexposure to carbon tetrachloride is an example.

The first step in treatment for inhalation poisoning is to remove the victim from the contaminated atmosphere. The rescuer must be protected with proper respiratory equipment or skin protection in order not to become a victim himself. (*See* Chapter 3, "Personal Protective Equipment.") If the toxic material is not obvious, look for labeled containers or sources from which the materials may have escaped or spilled. Many labels give instructions about what to do for overexposure to the contents. Medical assistance should be obtained promptly, but the victim should not be left alone. It is almost always possible to get additional help by shouting or attracting attention.

If the victim is unconscious, make sure his tongue, dentures, or vomit do not clog his breathing passage. Keep him warm. If he is conscious, have him lie down; if he breathes with difficulty, a sitting position may be better. If breathing has stopped, artificial respiration must be started at once. (*See* Chapter 1, "First Aid.") Mouth to mouth resuscitation is the best method, but if this cannot be applied use another method. Do not delay for even a fraction of a minute! Watch for symptoms of shock (*see* Chapter 1). Oxygen should be administered only by someone thoroughly trained in its use.

The preceding series of steps should be followed as closely as possible if some variation is required. The **Number One Safety Rule** is: *never work alone in a laboratory or in any isolated location. There is always the possibility of an accident.*

Exposure by Skin Contact

Skin contact is the most frequent method of exposure to most toxic chemicals. However, it does not cause as many fatalities or major poisonings as inhalation. Nevertheless, it produces much discomfort, inconvenience, and time lost. It can be fatal. Some chemicals penetrate the skin rapidly, are rapidly absorbed and transported by the blood, and cause systemic poisoning (upset of body systems). Tetraethyllead (the source of lead in gasoline) is an example. Other chemicals, such as strong acids or bases, may have serious local effects at the point of contact. Some chemicals act both ways. Eyes may be damaged not only by splashing chemicals into them but also by absorption of water-soluble gases such as ammonia or hydrogen chloride on the moist surface of the eyeball. Surface absorption can then cause serious damage even though the gas may not have been inhaled.

Skin contact damage can be seriously increased by clothing which has been wetted by the chemical. The wet clothing prolongs the skin's exposure to a high concentration. Contaminated clothing should be removed and the affected area flushed with water immediately, regardless of considerations of modesty.

The local effect may cause nothing more than a slight reddening of the skin or mild dehydration that causes skin cracks and roughness. However, it may cause major destruction of tissue. If the material is absorbed, the result can range from mild poisoning to death. Industrial dermatitis is often the result of prolonged low-level surface contact with an irritating substance.

Table II. Threshold Limit Values

A partial list of recommended values as adopted by the American Conference of Governmental Industrial Hygienists for 1973. Because of a wide variation in individual susceptibility, occasional exposure of an individual at, or even below, the threshold limit may cause discomfort, aggravation of a pre-existing condition, or occupational illness. These limits should be used as guides for control of health hazards and should not be regarded as limits between safe and dangerous concentrations. "C" listings are ceiling values not to be exceeded even when concentrations are determined on a time weighted average as may sometimes be done for substances not given a "C" listing. The ACGIH has established bases for allowing infrequent excursions in modest degree above the TLV if counterbalanced by concentration samplings adequately below the upper value for substances not carrying a "C" listing when a time-weighted average determination instead of constant monitoring is used. If the slightest doubt exists, refer to the publication "Threshold Limit Values for Chemical Substances and Physical Agents in the Workroom Environment with Intended Changes for 1973." Copies may be obtained from the Secretary-Treasurer, ACGIH, P.O. Box 1937, Cincinnati, Ohio 45201. An asterisk (*) indicates a notice of intended change. Ppm means parts of vapor or gas per million parts of air by volume at 25°C and 760 mm Hg pressure. MgM³ means approximate milligrams of particulate per cubic meter of air.

(Continued on page 38)

Table II.   Continued

Substance	ppm	Mg/M³	Substance	ppm	Mg/M³
Acetaldehyde	100	180	Chlorodiphenyl		
Acetic acid	10	25	(54% chlorine)—		
Acetic anhydride	5	20	skin	—	0.5
Acetone	1,000	2,400	* Chloroform	25	120
Acrolein	0.1	0.25	Chromic acid and		
Acrylonitrile—skin	20	45	chromates		
Allyl alcohol—skin	2	3	(as CrO₃)	—	0.1
Allyl chloride	1	5	Cobalt, metal fume		
Ammonia	25	18	and dust	—	0.1
n-Amyl acetate	100	525	Cresol (all iso-		
Aniline—skin	5	19	mers)—skin	5	22
Antimony and com-			Cyanide		
pounds (as Sb)	—	0.5	(as CN)—skin	—	5
Arsenic and com-			Cyclohexane	300	1,050
pounds (as As)	—	0.5	Cyclohexanol	50	200
Barium (soluble			Cyclohexanone	50	200
compounds)	—	0.5	2,4-D	—	10
C Benzene			DDT	—	1
(benzol)—skin	25	80	C o-Dichlorobenzene	50	300
Bromine	0.1	0.7	p-Dichlorobenzene	75	450
Butadiene			1,1-Dichloroethane	200	320
(1,3–butadiene)	1,000	2,200	1,2-Dichloroethane	50	200
2-Butanone	200	590	Diethylamine	25	75
2-Butoxyethanol			Dimethylaniline		
(Butyl Cello-			(N-dimethyl-		
solve)—skin	50	240	aniline)—skin	5	25
Butyl acetate			Dinitrotoluene—		
(n-butyl acetate)	150	710	skin	—	1.5
Butyl alcohol	100	300	* Dioxane (dieth-		
tert-Butyl alcohol	100	300	ylene diox-		
C Butylamine—skin	5	15	ide)—skin	50	180
* Cadmium oxide			Ethanolamine	3	6
fume (as Cd)	—	0.05	Ethyl acetate	400	1,400
Carbon dioxide	5,000	9,000	Ethyl alcohol		
Carbon disulfide—			(ethanol)	1,000	1,900
skin	20	60	Ethylamine	10	18
Carbon monoxide	50	55	Ethylbenzene	100	435
Carbon tetrachlo-			Ethyl bromide	200	890
ride—skin	10	65	Ethyl chloride	1,000	2,600
Chlorine	1	3	Ethyl ether	400	1,200
Chlorobenzene			Ethyl fomate	100	300
(monochloro-			Ethylene oxide	50	90
benzene)	75	350	Fluoride (as F)	—	2.5
			Fluorine	1	2

Substance	ppm	Mg/M³	Substance	ppm	Mg/M³
C Formaldehyde	2	3	2-Nitropropane	25	90
Furfural—skin	5	20	Octane	400	1,900
Heptane (*n*-heptane)	500	2,000	Oil mist (particulate)	—	5
Hexane (*n*-hexane)	500	1,800	Ozone	0.1	0.2
Hydrogen bromide	3	10	Parathion—skin	—	0.1
C Hydrogen chloride	5	7	Pentachloronaph-		
Hydrogen			thalene—skin	—	0.5
cyanide—skin	10	11	Pentachlorophenol—		
Hydrogen sulfide	10	15	skin	—	0.5
C Iodine	0.1	1	2-Pentanone	200	700
* Iron oxide fume	—	5	Perchloroethylene	100	670
Lead, inorg.,			Phenol—skin	5	19
fumes and dusts	—	0.15	Phosdrin (Mevin-		
Lindane	—	0.5	phos)—skin	0.01	0.1
LPG (Liquified			Phosgene (carbonyl		
petroleum gas)	1,000	1,800	chloride)	0.1	0.4
Magnesium oxide			Phosphine	0.3	0.4
fume	—	10	Phosphorus (yellow)	—	0.1
Malathion—skin	—	10	Phosphorus		
C Manganese and com-			pentachloride	—	1
pounds (as Mn)	—	5	Phosphorus		
Mercury (Alkyl			pentasulfide	—	1
compounds)—			Phosphorus		
skin	0.001	0.01	trichloride	0.5	3
Mercury (All forms			*n*-Propyl acetate	200	840
except alkyl)	—	0.05	Propyl alcohol	200	500
Methyl acetate	200	610	Selenium com-		
Methyl alcohol			pounds (as Se)	—	0.2
(methanol)	200	260	Silver, metal, and		
Methyl chloride	100	210	soluble com-		
Methylcyclohexane	500	2,000	pounds	—	0.01
Methyl formate	100	250	Styrene (monomer)		
* Methylene chloride			(phenylethylene)	100	420
(dichloro-			Sulfur dioxide	5	13
methane)	250	890	1,1,2,2-tetrachloro-		
Naphtha (coal tar)	100	400	ethane—skin	5	35
Naphthalene	10	50	Tetraethyllead		
Nicotine—skin	0.075	0.5	(Pb)—skin	—	0.100
Nitrobenzene—skin	1	5	Tetrahydrofuran	200	590
Nitroethane	100	310	Toluene (toluol)	100	375
			Xylene (xylol)	100	435

"Skin" Notation. Listed substances followed by the designation "skin" refer to the potential contribution to the overall exposure by the cutaneous route including mucous membranes and eye, either by airborne, or more particularly, by direct contact with the substance. This attention calling designation is intended to suggest appropriate measures for the prevention of cutaneous absorption so that the threshold limit is not invalidated. For Mixtures: *See* the ACGIH reference.

The treatment for skin contact with poisons or irritating substances is quick washing of the affected area with large amounts of water and then gently cleansing the area with soap and water. If the material clings tightly to the skin, care must be exercised in removing it with an organic solvent. The solvent may increase absorption of a substance that might not otherwise be absorbed. A bland medicinal oil such as paraffin oil is probably best if something other than soap and water is necessary. An eye wash fountain and shower should be available for rapid washing operations. Eye washing should be thorough. The victim should roll his eyes in all directions while holding the lids open. Eye washing should be continued for 15 minutes. If large areas of skin are affected or appreciable areas of clothing are contaminated, the safety shower is required for an adequate, rapid wash. Clothing can be removed under the running shower to prevent extending the skin exposure.

A solution of sodium bicarbonate should be available for treating strong acid exposures; a strong base should never be used for this purpose. Usually, the best treatment of strong base exposure is immediate washing with large amounts of water.

A severe case of contact poisoning or local tissue damage requires the immediate attention of a physician. All other cases, except for mild localized irritations, should be examined by a physician.

Exposure by Swallowing

Laboratory workers seldom swallow (ingest) toxic chemicals. When a toxic substance is swallowed, it must be removed or made harmless as quickly as possible. The old practice of using one's mouth as a vacuum source for a pipet should be completely prohibited. A swallowed chemical may cause only minor irritation to skin but may destroy the soft linings of the mouth, throat, and stomach walls. These surfaces rapidly absorb some chemicals which have little tendency to penetrate the outer skin.

Small amounts of chemicals may be swallowed when moistening one's fingers to flip pages in a notebook, licking labels, smoking, eating, or drinking in the laboratory. These practices should be completely avoided in chemical laboratories and other areas where toxic chemicals may be used. Containers used for coffee should never be stored in laboratory refrigerators or cold boxes. Before smoking or eating or drinking anything your hands should be washed. Make certain there is no possibility of swallowing any harmful substance. Failure to take these precautions can result in serious cumulative effects from repeated swallowing of tiny amounts of harmful chemicals. Never taste a chemical for identification.

It is sometimes difficult to determine the poison and amount that has been swallowed. The victim may be frightened, incoherent, or unconscious. The substance that is swallowed must be determined quickly; often a poison requires a specific type of treatment. The physician or the local poison treatment center will be badly handicapped without this information. If the substance came from a container, directions for counteracting the poison may be on the label. If the victim is conscious, he should be made to vomit (for important exceptions to this rule, *see Poisoning* in Chapter 1, "First Aid") by giving him a large amount of water to drink and sticking a finger down his throat. Certain emetics (solutions which cause vomiting) may be administered. If he is unconscious do not give him anything to drink. A physician should be called at once and told what poison was swallowed so that he can contact the local poison information center for instructions. Keep the victim lying down, warm, and comfortable. Watch for symptoms of shock that may need prompt treatment.

General Precautions

A chemical technician should not wait until an emergency arises to learn what procedures should be followed during a crisis. He should be familiar with emergency alarm systems to summon help. He should know how to call the nurse, physician, ambulance, emergency safety squad, or someone who knows the proper procedures. He should know the location of emergency equipment and be familiar with basic First Aid. Knowing the specific hazards and how properly to handle the particular chemicals being used are the marks of a good technician. No shortcoming of the chemical technician's performance reflects more seriously on his desirability than carelessness.

Handling Concentrated Acids

Reagents that are usually used to dissolve inorganic materials are hydrochloric acid (HCl), nitric acid (HNO₃), a mixture of three parts of concentrated

hydrochloric acid and one part of concentrated nitric acid (aqua regia), sulfuric acid (H_2SO_4), hydriodic acid (HI), hydrofluoric acid (HF), perchloric acid ($HClO_4$), and sodium hydroxide (NaOH). These reagents may be used in either concentrated or dilute forms. **CAUTION: Remember that when mixing acids with water, always add the acid to the water.** Adding water to acid is dangerous because the heat developed may cause the acid to boil and spatter out of the container.

Hydrochloric Acid

Concentrated HCl is 12.0*M*. The fumes of HCl are very irritating to the nose and eyes. Whenever you use this acid in concentrated form, work in a hood. HCl can be used to dissolve many metals and metal ores in the carbonate and oxide forms.

Nitric Acid

Concentrated HNO_3 is about 15.7*M*. The freshly prepared acid is colorless, but after standing the acid becomes yellow from the nitrogen dioxide (NO_2) that forms as the acid slowly decomposes. This does not spoil the reagent, and it can still be used. Nitric acid is a very powerful acid and should be handled with extreme care. Its fumes do not irritate the eyes and nose like HCl, but splashes of nitric acid on the skin will turn it yellow in a few minutes and after a few days the skin will peel off. If you get a few drops of HNO_3 on you, it generally will not burn immediately; consequently it is easy to get burned and not know it for several minutes. If you suspect that you have spilled this acid on you, wash the suspected area immediately. Nitric acid will dissolve most metals not dissolved by HCl except iron and aluminum. It can be used for dissolving alloys and sulfide ores. Aqua regia (HNO_3–HCl, 1:3) is required to dissolve alloy steels, gold, and platinum.

Sulfuric Acid

Concentrated H_2SO_4 is 17.8M. A bottle of this acid is surprisingly heavy and many serious accidents have occurred when students have tried to pick up a bottle and then let it drop. Always use an acid carrier. H_2SO_4 can be used to dissolve the oxides of aluminum and titanium.

Acid carriers

Sulfuric acid is a powerful dehydrating agent, and it will char organic material. If it gets on you, it immediately causes burns, and the skin will start to turn gray. Before you can wash the affected area, the top layer of skin will begin peeling and you will have a very sore area for a few days. A weak base solution such as sodium bicarbonate ($NaHCO_3$) should be applied to the affected area immediately. Exercise great care with this acid. There are probably more serious laboratory accidents with this acid than any other except for perchloric acid. You have been warned not to pour water into acid, and this is particularly important for concentrated sulfuric acid. The

following true story will serve as an example of how easy it is to get hurt.

A girl was working at a laboratory bench with two other students. She finished before the others, cleaned up her equipment and was about to leave when one of the other students asked her if she would wash out a few test tubes. She went to the sink, turned on the water and put the test tubes under the faucet. Immediately the material from one of the tubes blew back into her face. One of the tubes had concentrated H_2SO_4 in it, and when the water hit it, the heat generated was so great that a virtual jet of acid was shot into her face. Fortunately she had on safety glasses and her eyes were saved. She lost most of the top layer of skin on her face, and she was not a pretty sight for several days.

Remember this the next time you handle concentrated acids.

Hydriodic Acid

Concentrated HI is 7.6M. The pure material is colorless but with age the acid will be colored reddish-brown because of free iodine. This acid is used to dissolve mercury, tin, arsenic, calcium fluoride, CaF_2, and the barium, strontium, and lead sulfates which are very difficult to dissolve.

Hydrofluoric Acid

Concentrated HF is 27.6M. Analytically it is most widely used to dissolve silicon oxide, SiO_2; this compound, also called silica, is a major constituent of sand, glass, and many rocks. **CAUTION: Be very careful when handling hydrofluoric acid. Always wear plastic or rubber gloves.** Burns from this acid are very painful and last for several weeks. The most dangerous burns from this acid occur if some of the acid gets under the fingernails. If this happens, see a physician immediately and have a calcium gluconate injection administered to the affected fingers. Calcium reacts with the HF to form insoluble and un-reactive CaF_2. Unless the calcium is added the HF will burn all the way to the bone where it will react with the calcium in the bone and become neu-tralized. Spills of HF present numerous hazards and must be handled immedi-ately. Hydrofluoric acid dissolves glass very rapidly and is never stored in a glass bottle (unless paraffin-lined). Reactions involving it are never carried out in glass containers. Always use plastic ware or platinum vessels when using this acid. The reaction with glass containers is rep-resented by:

$$SiO_2 + 4\ HF \rightarrow 2\ H_2O + SiF_4 \uparrow$$

silica in glass   silicon tetra-fluoride (a gas)

A similar reaction can be used to etch glass.

Perchloric Acid

Perchloric acid generally is sold in pint bottles of either 60% ($9.9M$) or 72% ($11.9M$) acid. When this acid is cold and dilute, it acts like any other strong acid. It is colorless and odorless and vapors from it do not affect the nose or eyes. However, if this acid becomes hot and concentrated it is a very powerful oxidizing agent. This acid is very effec-tive in dissolving chrome steels.

Sodium Hydroxide

Bases are seldom used to dissolve ma-terials. However, the amphoteric metals —aluminum, tin, lead, zinc, and chro-mium(III)—are easily dissolved by NaOH. A boric acid solution is used to treat burns caused by bases.

Other Toxic Chemicals

Every chemical, whether found in some toxicity listing or not, should be considered toxic until proved non-toxic. The old saying "familiarity breeds con-tempt" is rarely more applicable. Be-cause they are used so much, many solvents are considered harmless when they are actually very dangerous. Ben-zene is an excellent example since its TLV is considered to be only 25 ppm. In many reactions toluene or xylene could be substituted for benzene and they have a TLV that is eight times higher.

There is constant research on the tox-icity of specific chemicals. Nevertheless, it is sometimes impossible to find refer-ences about the hazards of a specific substance. The most comprehensive and readily available list of hazardous air-borne substances is published by the American Conference of Governmental Industrial Hygienists.(6) In addition to the many substances included in the ACGIH booklet, some other common solvents should be considered toxic:

    diethylene glycol
    ethylene glycol
    glycol ethers and esters
    halogenated hydrocarbons
    esters of organic acids
    amines

Solvent spills on skin or clothing must be avoided as carefully as inhalation of vapors.

When considering skin or ingestion effects of poisons that are more normally inhaled and on which specific informa-tion is not available, make conservative estimates about the poisoning on the basis of what is known. There are many correlations between toxicity and chemical composition or structure. Therefore analogies are important. Certain types of compounds and certain reactive groups in the molecule allow some prediction to be made about their poisonous character.

The following compounds and reactive groups are given without any attempt to differentiate moderate from severe toxicity and without regard to how the substance gets into the body.

Special Toxic Groups

Some organic compounds are potentially toxic because they contain certain functional groups that are called *toxophores*. The list of toxophores includes:

$$\ce{>CO}, \quad \ce{>S}, \quad \ce{>C=C<},$$

$$\ce{-N<^O_O}, \quad \ce{N=C<}, \quad \ce{As-}$$

If an organic molecule containing a toxophore also includes any member of a second set of functional groups called *autotoxics,* a poisonous compound is produced. The list of autotoxics includes:

$$\ce{>O}, \quad \ce{-Cl}, \quad \ce{-NH_2}, \quad \text{(phenyl)},$$

$$\text{(phenyl)}\ce{OH}, \quad \ce{-CH_3}, \quad \ce{-C_2H_5}$$

It should be noted that some substances become even more dangerous in the presence of a second substance. When a nitro- compound or carbon tetrachloride has been taken into the body, the poisoning effect is much worse if alcohol has been consumed at about the same time.

Local effects to the skin may be dehydration, dissolution of protective skin constituents, or precipitation of protein. Chromic acid and chromates are specific skin poisons. Sulfur (thiols, sulfides, etc.) increases the capacity of a substance to penetrate skin.

Nitrites and R–N–NO compounds absorbed in the system cause blood vessels to dilate. The changes in blood pressure that result affect the respiratory balance. Aromatic amines, nitroso-, and nitro- compounds attack blood cells. Many compounds have a mild to severe narcotic effect because they attack parts of the nervous system. This is true of the hydrocarbons, alcohols, esters (particularly as molecular weight increases), and chlorinated hydrocarbons. Certain aromatic amines and azo- compounds can start cancers.

When certain groups or atoms are substituted into a molecule, they increase its irritating capacity or toxicity. This is true of nitro- groups, chlorine when it is introduced into a nitrated compound, and double bonds. Introducing chlorine into a compound will sometimes reduce one undesirable property and simultaneously increase another one like toxicity. Fluorine-containing compounds must usually be considered poisonous, particularly if the compound also contains phosphorus. Certain phosphate and pyrophosphate esters are such powerful poisons that they are used as insecticide sprays.

Additional chemicals or families that should be considered potentially toxic include:

white phosphorus
heavy metals and their vapors
oxalic acid
soluble metallic cyanides
isocyanic acid esters

chlorates of sodium, potassium, and
   ammonium
acetylides
compounds of:

lead	arsenic
mercury	antimony
silver	copper
zinc	selenium
beryllium	thallium
cadmium	

aldehydes
carbonyls (iron, cobalt, nickel)
oxides of nitrogen
cyanogen
phenols, phenolic compounds
halogens
hydrogen halides
boron halides
nitriles
nitro- compounds
halogenated hydrocarbons
organic and inorganic sulfur compounds—thiols, mercaptans, thio-
   esters, sulfides, disulfides, etc.
fumigants
pesticides
alkaloids
some glucosides

It seems appropriate to close by nam-
ing five particular dangerously poisonous
vapors.

Carbon Monoxide (TLV 50 ppm)

Carbon monoxide is very poisonous
and will cause death if inhaled for just
a few minutes. It gives no warning by
smell or taste, and a victim's first symptoms are headache, dizziness, and general weakness. By this time irreversible
damage may have been done. At an atmospheric concentration of less than
0.05%, a human can lose consciousness
in a matter of minutes after the first
symptoms appear. At a concentration
of 0.1%, death can follow in minutes.
Pieters and Creyghton (7) report that
more accidents are caused by carbon
monoxide than all other toxic gases combined. This gas is readily produced
wherever incomplete combustion of a
carbon-containing substance takes place.
It is produced by internal combustion
engines, luminous flames, flames in contact with a cold surface, home barbecue
cookers, producer gas, etc. Every year
many deaths occur in garages, automo-
biles, and trailers or campers in which
the engine or a heater has been left running. The installation of unvented heaters in living quarters results in many
cases of carbon monoxide poisoning.

Mercury Vapor (TLV 0.1 mg/m³)

Most people handle mercury too cas-
ually, considering its toxicity. It is widely
used in glass instruments that are easily
broken so some exposure is inevitable.
But anyone who works frequently or
regularly with mercury should study its
hazards and safe handling. He should
also be under regular medical supervision.

The TLV for mercury vapor is 0.1
mg per cubic meter of air. One cubic
meter of air saturated with mercury vapor at 20°C contains about 15 mg.
Therefore air is dangerous at less than
1% saturation at room temperature.
When mercury is spilled, it must be collected scrupulously. (*see* Chapter 6,
"Special Hazards.") It should not be
handled in open containers or used as a
seal in equipment where any appreciable
surface is exposed. The distilling of mercury or its use around mercury diffusion
vacuum pumps should be left to carefully
trained experts. It should never be
warmed except in a closed system from
which vapors can be safely exhausted.
Smoking and eating near mercury should
be prohibited. Carelessness cannot be
tolerated, and personal cleanliness is im-

perative. Special clothing that is free of fuzz, seams, cuffs, pockets, and ornamentation is highly recommended. It should be laundered frequently by an industrial laundry.

Extra precautions are in order because mercury has a cumulative effect. The sum of many insignificant exposures can lead to serious poisoning. The symptoms are emotional and physical instability and various types of muscular tremor affecting the extremities, eyelids, and even the tongue. An odd taste and sore mouth and gums, diarrhea, and reduced appetite follow. A good reference for further information on handling mercury, written by Bidstrup,(8) also describes a special detecting soap which changes color when it contacts mercury.

Hydrogen Sulfide (TLV 10 ppm)

Hydrogen sulfide is frequently handled as though it is not poisonous. In fact, its TLV is only one-fifth that of carbon monoxide. It is especially dangerous—it has a detectable and characteristic odor, but the sense of smell is quickly dulled leading the victim to believe that the hazard no longer exists. Hydrogen sulfide poisons the blood and is a systemic poison with a variety of symptoms. Since it is water-soluble it also attacks the eyes causing painful and damaging reactions. Exposure to concentrations only a little above the TLV causes headache, nausea, dizziness, shallow breathing, and lowered blood pressure. Exposure to a concentration greater than 0.5% hydrogen sulfide in air results in loss of consciousness almost at once because the gas disables the respiratory control center. A victim of hydrogen sulfide poisoning needs help immediately. Nothing should delay getting him into an uncontaminated atmosphere, calling a physician and applying artificial respi-

ration if he is unconscious. Details are included in Chapter 1, "First Aid."

Carbon Tetrachloride (TLV 10 ppm)

In spite of its highly toxic nature, the careless use of carbon tet as a cleaning solvent and as a filler for pump-type fire extinguishers has generated an appalling hazard outside the laboratory. The ravages of this killer have been well documented, but it continues to be used widely and unwisely.

A few quotations from the National Safety Council's quarterly, *Family Safety,* (9) should be sufficiently illustrative.

A mother in New York City recently used half a cup of carbon tetrachloride to clean the upholstery of two chairs. That afternoon her seven-year-old daughter came home from school with a slight cold and sat on one of the chairs to watch television. Later the child became so violently ill that she was taken to a hospital where doctors found that her kidneys had stopped functioning and she was near death. The cause was inhaling the poisonous vapor of carbon tet. It took the efforts of a ten-man medical team and an artificial kidney to keep the child alive. She remained in the hospital six weeks until her kidneys healed.

A young man in New Jersey, after having a few drinks, cleaned his shoes in a closet with a little carbon tet. He became ill but did not enter a hospital until three days later. He died within four days and an autopsy confirmed that the culprit was carbon tet. His liver and lungs had been severely assaulted, and his kidneys had completely failed.

Carbon tet evaporates quickly when exposed to air, giving off vapor that is more poisonous than that of its cousin, chloroform. Only three thimblefuls of carbon tet will saturate the air to the danger point in an unventilated room ten feet square by ten feet high.

To make matters worse, the victim who inhales carbon tet usually has no warning *at the time* that he is being poisoned. Symptoms appear after the damage has been done. The carbon tet is taken from the lungs to the bloodstream and deposited in the fatty tissues of the body, primarily the brain and liver. A stricken person may become dizzy, nauseated, jaundiced, and begins to cough. Often he suspects he is coming down with the flu.

For some unknown reason, having alcohol and carbon tet in the body simultaneously makes a person many times more susceptible to poisoning.

So-called adequate ventilation to make the use of carbon tet safe is rarely if ever achieved indoors except inside a properly operating laboratory exhaust hood.

Medical help in a hospital is a must for serious poisoning by carbon tet.

Benzene (TLV 25 ppm)

Benzene, a popular solvent, has been mentioned as a poison. It is included here for emphasis since it is widely used in the laboratory and in industrial processes. Its odor is faint and not disagreeable. Symptoms may develop only after considerable damage has been done, so great care must be exercised to keep exposure near zero. Much of the damage from chronic benzene poisoning is irreversible and may get worse after exposure ceases. It is a systemic poison and concentrates much of its attack on bone marrow, thereby destroying blood cells and preventing their regeneration. Brief inhalation of high concentrations may result in unconsciousness and death. Chronic poisoning symptoms include headache, weakness, dizziness, and unexpected bleeding at nose and mouth. An anemic appearance and urinary upsets may be noticeable. There are medical tests which can identify benzene poisoning, and these should be used in all suspected cases. The body is not equipped to rid itself of benzene easily.

It must be clear that handling chemicals without sufficient knowledge of their characteristics is not only dangerous but professionally inexcusable. It is essential that a laboratory technician always work with regard to his safety and the safety of others. Clearly, many substances are toxic, can impair health, and are explosion or fire hazards.

The Merck Index

The Merck Index is a valuable resource when searching out these characteristics of a particular chemical substance. It is the one publication which gives levels of toxicity, trade names, information on structure, and the general characteristics of chemicals and drugs, and the health hazards involved in working with the substances.

Employers' awareness of the economic as well as the human benefits of a well-planned safety program has paved the way for safe working conditions. Proper use of the excellent laboratory and protective equipment now available coupled with sufficient knowledge of the hazards inherent in various chemicals, can virtually eliminate illness and disability caused by toxic chemicals.

Literature Cited

1. Cummings, John N., "Heavy Metals and the Brain," Charles C. Thomas, Springfield, 1959.
2. Steere, Norman V., "Mercury Vapor Hazards and Control Measures," *J. Chem. Educ.* (1965) **41**, A529.
3. Manufacturing Chemists' Association, Inc., "Health Factors in Safe Handling of Chemicals," *Safety Guide SG-1* (1960) 1825 Connecticut Ave., N.W., Washington, D.C.
4. *California Safety News,* California State Department of Industrial Relations, Div. of Industrial Safety, 445 Golden Gate Ave., San Francisco, (1964) **48** (3).
5. West, Irma, "Public Health Problems are Created by Pesticides," *Calif. Health* (1965) **23,** 1.
6. American Conference of Governmental Industrial Hygienists, Cincinnati, Ohio.
7. Pieters and Creyghton, "Safety in the Chemical Laboratory," Butterworth, London, 1957.
8. Bidstrup, P. Lesley, "Toxicity of Mercury and Its Compounds," Elsevier, Amsterdam, 1964.
9. National Safety Council, *Family Safety,* 425 N. Michigan Ave., Chicago, Ill.

Special Hazards

All chemicals should be carefully handled. However, sodium, mercury, and organic peroxides present special hazards. An accident involving any of these substances is usually dangerous and requires particular attention.

Sodium

Sodium reacts readily and vigorously with water. If a small pellet of sodium is added to water, the metal reacts quickly, melts, skitters about on the surface as a molten ball, and rapidly disappears. Gas bubbles appear as the sodium reacts, and the gas frequently catches fire. The water in which the sodium was placed becomes more alkaline than it was originally. The chemical equation for this reaction is represented by:

$$2Na + 2H_2O \rightarrow 2Na^+ + 2OH^- + H_2$$

A large quantity of sodium is used in the chemical industry, and it is very important that chemical technicians be able to handle this element safely. Sodium reacts so rapidly with oxygen and water that it must be stored under kerosene or mineral oil. Even then some oxide or peroxide may form on the surface of the metal. When you look for sodium metal in a chemistry laboratory, you will probably find a bottle or can containing some lumps or cubes of white or grayish-white material immersed in oil or kerosene.

There will be no luster, and the surface coating will appear to be brittle. These physical properties are not characteristic of metal. The material observed on the surface must be an oxide or peroxide of the reactive sodium. Sodium metal is very soft and can be cut with a knife. It can be easily pressed into wire by

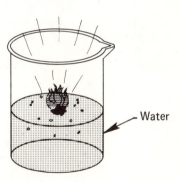
Water

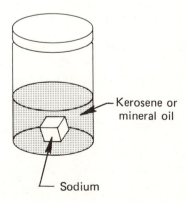
Kerosene or mineral oil

Sodium

merely pushing a lump of the metal through a die with a small hole which is held in a sodium wire press.

To make sodium wire, use tongs to remove a sodium metal lump from the kerosene bath and place it on a dry paper towel. The kerosene is absorbed by the paper. The layer of peroxide must be removed with a knife so that the silvery-looking metal is exposed. After the proper weight of material is obtained, it is put into a sodium press and converted to wire. Because the wire has a large amount of surface and will react rapidly with oxygen, the wire should be used immediately or placed in an inert solvent.

Clean the sodium press and dispose of the material that was removed from the original lump of sodium. As you know, it is dangerous to add sodium to water. Thus, small amounts of sodium residue and objects (such as the die in the press) that have sodium on them should be placed in ethyl alcohol which reacts much more slowly with sodium than does water. When hydrogen is no longer produced, the sodium metal has reacted, and it is safe to discard the alcohol solution. The die for the sodium press should be rinsed with alcohol and returned to its proper place. If sodium metal is exposed to air, it reacts with oxygen and the resulting peroxide is very deliquescent (attracts and absorbs water) and a very concentrated solution of sodium hydroxide results. This is highly corrosive and hazardous.

MCA Chemical Safety Data Sheet SD-47 is an excellent reference for the technician planning to work with sodium.

Mercury

The vapor coming from liquid mercury is extremely poisonous, about 100 times more poisonous than hydrogen cyanide (*see* Chapter 5, "Toxic Chemicals"). Also, unlike the cyanide, mer-

cury is a cumulative poison. This means that it remains in the body for long periods of time, and a harmful dose can develop after repeated exposures to small amounts of the material. The safe handling of mercury requires forethought, careful working habits, and prevention of situations that permit large mercury surfaces to be exposed to the atmosphere.

Forethought starts with the design and maintenance of the laboratory. The bench top and floors should be smooth and free of cracks. Whenever mercury is being handled where spills could occur, all work should be done over shallow pans or glass trays. This work should be done in fume hoods where spilled mercury can be completely removed and vapor can be easily vented. When mercury spills, it breaks into tiny droplets and bounces or rolls around. It gets into crevices, and tiny droplets can even stick to slanted or vertical surfaces. If mercury droplets lodge near heat sources, vapor is generated more rapidly than at room temperature.

All spilled mercury must be collected and returned to a closed container. Most of the mercury from a major spill can be carefully brushed into a special mercury collector or dust pan and transferred to a suitable container for safe disposal. The affected area can then be

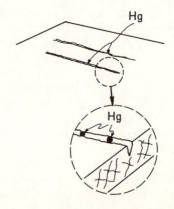

dusted with powdered sulfur or wet, fine sawdust and thoroughly cleaned again.

There will almost always be tiny droplets still clinging to the floor, under fixed furniture or in cracks. These can be collected by using a glass tube with a capillary tip connected through a collecting bottle to an aspirator. This is used as a fine-tipped vacuum cleaner that can get into cracks, crevices, and corners. It will collect most of the remaining droplets even though some are so small that they are very difficult to see. It will also collect dirt and dust but these will help to act as carriers of some of the fine droplets. The capillary tip may require frequent cleaning to prevent clogging.

The last traces of spilled mercury are nearly impossible to recover. Therefore, some dependence must still be placed upon good overall ventilation to avoid any buildup of vapor even after a thorough cleaning job. This is why such serious attention must be given to avoiding mercury spills in the first place. A mercury spill may be large if it occurs during a transfer operation. The liquid metal is so heavy that a technician who is not familiar with the rapidly shifting weight distribution in any mercury container should be especially alert. It is necessary to keep a firm grip on mercury flasks or other mercury containers and to make transfers very slowly. Otherwise, an unexpected rush of the liquid over the lip of the pouring spout can be very surprising and a time-consuming cleaning job may result.

Some laboratories have special boxes or well-ventilated, smooth-surfaced rooms in which mercury transfers or other similar operations are performed. A well-designed area emphasizes the need for care, makes cleaning easier, and provides ventilation to protect the individual doing the cleaning.

Any questions concerning further precautions should be discussed with your supervisor. (A good reference for a more detailed discussion is N. V. Steere, "Mercury Vapor Hazards and Control Measures," *Journal of Chemical Education* (July 1965) **42** (7)A529.

Peroxides

Peroxides are a special source of hazard in ethers and ether storage. The long history of explosions caused by ethers that have been stored too long has not been given as much publicity and attention as it deserves. One such accident caused the death of a chemist who was opening a pint bottle of isopropyl ether (MCA Case History No. 603).

Ethers form peroxides upon standing for periods of weeks or months, especially in the presence of air (oxygen). Peroxides are non-volatile and tend to concentrate when ether is evaporated.

$$RCH_2OR + O_2 \rightarrow \underset{\underset{O-OH}{|}}{RCHOR}$$

a peroxide

Peroxides will decompose upon being heated, and when the decomposition reaction is started, it goes so rapidly that

an explosion (detonation) likely results. Sometimes an ether container in which peroxides have formed will explode simply from being jarred or struck when at room temperature.

Ethyl ether is widely used in laboratories as an extraction solvent. It is necessary to test ethyl ether for peroxides that may be present in new as well as in older supplies. **Diisopropyl ether has a very strong tendency to form dangerous amounts of peroxides.** Tetrahydrofuran and dioxane, two commonly used solvents, also tend to form dangerous peroxides upon long exposure to oxygen.

There are convenient tests for the presence of peroxides. Peroxides will oxidize I^- to free iodine (I_2) which produces its characteristic color. A useful procedure for this test is to place 10 ml of the ether in a clean flask or graduated cylinder which has been rinsed with a sample of the ether. Add 1 ml of freshly prepared 10% aqueous potassium iodide (KI) solution and mix thoroughly. Allow the mixture to stand for one minute and examine it for the presence of a light yellow color due to free iodine. The presence of the color indicates that peroxides are in the ether. Fe (II) can also be oxidized to Fe (III) with the peroxides and provides the basis of another test method.

If peroxides are found in an ether sample, they may be removed by passing the ether through a column of activated alumina (Al_2O_3) or by washing it in a separatory funnel with a dilute aqueous ferrous sulfate ($FeSO_4$) solution.

The presence of water or metallic iron and protection from light and heat slow down the formation of peroxides. Safety rules should be observed even if ethers are bought in metal containers and kept cold. This is particularly important for absolute (anhydrous) ethers.

1. Purchase ethers in small containers which are adequate for only a few weeks' supply.

2. Date all containers on the labels to show when they were received. Remember that they may have been stored at the supply house for some time before being shipped to your storeroom!

3. Establish time limits for keeping opened containers of all ethers. Another time schedule should be used for testing or discarding ethers stored in unopened containers. These time limits depend upon the ether and whether it is anhydrous, inhibited, or kept in glass or metal containers.

4. Work behind sturdy safety shields, unless reagents are known to be peroxide free.

5. All technicians should be alerted to the hazards of ethers before handling them.

Two good references are:

1. MCA Chemical Safety Data Sheet SD-29 (for ethyl ether)

2. N. V. Steere, "Control of Hazards from Peroxides in Ether," *Journal of Chemical Education,* **41** (8) A 575 (August 1964) (for ethers as a class)

Cryogenics

Working with materials at very low temperatures, *i.e.,* $-75°C$ or lower, poses a special hazard. Cryogenic technology deals with liquified gases which are capable of enormous volume increase in going from liquid to vapor. Very low temperature work should never be undertaken without careful schooling in the hazards and necessary precautions. The extraordinary hazards of flammability, generation of high pressure, changes in physical characteristics of structural materials, and vulnerability of human tissue to extreme cold call for absolute safety. An excellent introduction is an article by E. W. Spencer from the *Journal of*

Chemical Education, reprinted in N. V. Steere (Ed.), "Safety in the Chemical Laboratory," p. 79-82, American Chemical Society, Division of Chemical Education, Easton, Pa., 1967.

Unattended Operations

Safety rules for operations not constantly attended must be tailored to all possible hazards. Leaving a damp, inert sample overnight in a thermostatically controlled oven at 110°C to dry poses nearly no hazard. However, leaving an unattended reaction under reflux or a prolonged distillation demands safety controls. Heating units may fail, the flow of cooling water may be interrupted, unacceptable levels of bumping or foaming or overheating may develop, and equipment may break or separate. It is not enough to ask a watchman to drop by once in a while to make sure all is well. The variety of hazards is too great and the numbers and kinds of monitoring and safety devices too many to include much more than a serious warning here.

Without a lot of experience it is unsafe to assume that you can foresee all the unsafe possibilities. Study some references (*1*) until you feel confident that with your safety precautions there is no appreciable chance of an accident. Finally make sure that your supervisor knows exactly what you are doing and approves your course of protection.

Refrigerators

Do not store food of any kind in a laboratory refrigerator and do not use a laboratory refrigerator or ice cubes from it to cool beverages. Containers of flammable gases or liquids must never be stored in conventional refrigerators. Special explosion-proof refrigerators are made with no exposed electrical contacts. Conventional refrigerators which have been converted by putting all contacts outside the box and eliminating light bulbs and sockets are safe to use. It is easy to underestimate this hazard, and there have been many serious accidents and explosions as a result.

Miscellaneous Special Hazards

1. *Vacuum* can cause implosions by the collapse of a vacuum desiccator, a Dewar flask, or a reaction or distillation setup. Dewar flasks, desiccators, and some vessels can often be protected by wrapping with special adhesive tape. This may not prevent collapse, but it eliminates flying glass fragments as a result. Metal containers, wire mesh cages, and sturdy protective shields are other useful safety devices.

2. *Excessive pressure* in ordinary laboratory glassware can cause serious explosions. Pressure reactions must be carried out in equipment designed to withstand safely the maximum pressure to which the equipment might be subjected. Overpressure safety releases may be necessary.

3. *Disposal practices* need careful thought. A jar labeled organic wastes can be a real trap. Waste disposal was mentioned in Chapter 2. When more hazardous materials than are described are involved, think carefully about the proper handling of any chemical destined for discard. There are good, inexpensive references (*2, 3*) available for study by anyone charged with disposal. Every laboratory should have someone responsible for safety in the disposal of chemicals.

4. *Auxiliary exits* are important components in the overall safety planning for any laboratory. Every laboratory should have an auxiliary exit for quick use if fire, fumes, or explosion blocks the normal path. This auxiliary path must not be blocked by equipment, file cabinets, or any other obstruction.

Literature Cited

1. Conlon, D. R., "Unattended Laboratory Operations," reprinted from *J. Chem. Ed.*, in Steere, N. V., Ed., "Safety in the Chemical Laboratory," p. 83-93, American Chemical Society, Division of Chemical Education, Easton, Pa., 1967.

2. Manufacturing Chemists Association, *Safety Guide SG-9*, 1825 Connecticut Ave., N.W., Washington, D.C.

3. Voeglein, J. F. Jr., "Storage and Disposal of Dangerous Chemicals," reprinted from *J. Chem. Ed.*, Steere, N. V., Ed., "Safety in the Chemical Laboratory," p. 72-75.

7

Radiation Hazards

Radiation from radioactive substances presents different hazards because it cannot be detected by human senses. It can be detected however with special instruments. The purpose of this chapter is to give you a working knowledge of the hazards of radiation and how to protect yourself and others around you from it.

Radioactive materials emit energy which has the power to damage living tissue. All radiation can do damage to the body if it is received in large enough doses. But within certain limits the damage can be repaired by the body so that there is no apparent effect. It is safe to work with radioactive substances if the proper limits of radiation are maintained in the working area.

Sources of Radiation

We must realize that radiation has always been a part of our environment. Before man used x-rays or radioactivity, there were many sources of natural radiation: cosmic radiation (from outer space); small amounts of potassium, uranium, thorium; and radioactive carbon in ordinary dirt, rocks, and food. Radioactive carbon formed by cosmic rays is part of every cell of our bodies. We shall see later that these natural sources are not harmful to man because they do not occur in concentrated forms.

The main difference between natural radiation and man-made radiation is that the man-made sources can be very strong.

The safety record of work involving atomic energy is excellent. Over a 15-year period, the thousands of people engaged in this work have developed very good safety records. The fatal accident rate is less than half of that of the best of American industry. Although 200 people have been killed in the program, only three were in accidents involving radiation. The others were killed in normal industrial accidents—fires, falls, electrocutions, motor vehicle accidents, and the like.

Basically, two types of exposure to radiation are possible (Figure 1); the radiation source can be external to the body, or it can be internally deposited by ingestion or entry through other openings in the body. A radiation overdose has several effects on the body.

Symptoms of Overexposure

Radiation sickness is produced by a massive overdose of penetrating external radiation. It causes nausea, vomiting, diarrhea, malaise, infection, and hemorrhage. If serious enough, it can cause death.

Radiation injuries occur from overdoses of less penetrating radiation that injures a small part of the body. Examples are burns, loss of hair, and skin lesions. This type of injury happens most often to the hands because contact is usually made with the hands. Radiation can also cause genetic damage.

Radioactive poisoning occurs when dangerous amounts of certain types of radioactive materials enter the body causing such diseases as anemia and cancer.

We realize that we can get into trouble with radiation by two entirely different means. One is from radiation coming at the body from the outside, and the other is from radioactive materials which have been taken into the body (Figure 1). To make our study easier, we will consider only external radiation hazards for the next part of our discussion. We will discuss the problem of the internal radiation hazards later.

External Radiation

External radiation comes from two sources:

1. Long-range, highly penetrating external radiation.

2. Short-range, less penetrating external radiation.

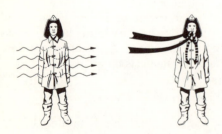

Figure 1.   *Types of radiation exposure*

External Radiation
Some radiation goes through the body like x-rays.

Internal Radiation
We can receive radiation by swallowing or breathing radioactive materials.

Long-range, highly penetrating external radiation is similar to x-rays and consists of pure energy with no mass. These rays originate from radioactive materials outside the body and come at the body like a shower of arrows or a beam of light. In order to simplify our problem, let us think of one ray at a time. Each ray is a bundle of energy. It penetrates the body to some depth before it does its damage and spends its energy.

Observe the man in Figure 2. Rays come at him from a radioactive source which is emitting penetrating radiation. Each one of the rays penetrates to a different depth in the body before it strikes a target, damaging a cell. A sizable proportion of the radiation passes through the man's body without ever touching him. These rays do no harm.

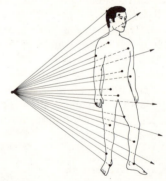

Figure 2.   *Results of exposure to a source of penetrating radiation*

At first this may seem strange. Most of us would agree that light rays cannot pass through a man's body without touching him. However, if we place our fingers over the lens of a strong flashlight, some light comes through our fingers. To explain this, recall that x-rays are like ordinary light rays but with a very short wavelength. If it were not for the fact that some radiation can pass through the body without touching it, we could not take x-ray pictures. For a chest x-ray process, the source of x-rays is pointed at the patient's back, and the film is placed over his chest. If his body stopped all of the radiation, none would reach the film. But some of the radiation does get through to the film. Generally, the more dense parts of the body absorb more x-rays than the less dense thus giving the x-ray picture.

Figure 3 shows what a sample of water might look like if we could magnify it with a microscope tremendously more powerful than any in existence.

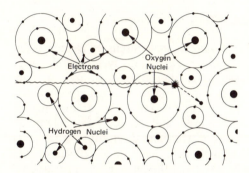

Figure 3. A diagram of water interacting with penetrating radiation

The atoms consist of two parts—a heavy, dense core called the nucleus which contains practically all of the weight, and tiny particles called electrons that occupy the space around the nucleus. The trouble with our illustration is that it is not at all to scale. If the nuclei of our atoms were as big as we have shown, the electrons would be hundreds of feet away. Since the electrons are relatively far away from the nucleus, then the nuclei of the various atoms must be quite far apart. In other words, any sample of matter contains much empty space on a microscopic level. Therefore, we see how it is possible for the ray to penetrate quite a distance into a substance before it hits anything.

Most of the rays from radioactive materials do not hit the nuclei. They do hit the electrons, which are moving about the nucleus, and the energy of the ray is spent. The removal of an electron from an atom or molecule is called ionization. Most radiation damage occurs because matter becomes ionized, and secondary chemical reactions occur.

The Roentgen

It is necessary to have some unit of measurement for radiation exposure to determine acceptable limits. The unit of measurement for penetrating external radiation exposure is the **roentgen,** abbreviated *r*. The milliroentgen, abbreviated as *mr,* is 1/1000 of a roentgen. Thus, we speak of a dose of radiation as so many r or so many mr.

Single Exposures

It is impossible to say how many roentgens it will take to kill any individual because we all vary in our resistance to any attack upon the body, whether it is by radiation, electricity, poison, injury, disease, etc. It is quite certain, however, that no human being could survive 1,000 r of total body radiation delivered in a short space of time. Both the total dose, the concentration, and the time are important considerations. The effect of 1,000 r of radiation delivered to the total body is by no means the same thing as 1,000 r delivered to a small portion of the body any more than a first degree burn of the palm of the hand is the same thing as a third degree burn of a large area of the body.

Generally, 24 hours or less is considered a short time. The ability of the body to withstand any injury is increased if the amount of injury given to the body

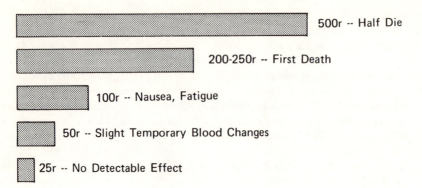

Figure 4.   Effects of external radiation for total body exposure within a 24-hour period

is spread over a longer period of time. Whiskey can be poison, but many people can drink an ounce of whiskey each evening before dinner over an extended period of time without apparent harm. If a person drank a fifth of whiskey in a few minutes, he could die of alcoholic poisoning because the body has not been given sufficient time to recover from the effect of the alcohol.

The radiation dose it takes to kill one individual is not a good measure of the fatal dose to others because of individual differences. The term used is the *median lethal dose,* or LD/50. This is the dose required to kill 50% of the subjects. The LD/50 for penetrating external radiation is about 500 r delivered to the total body in 24 hours or less. This means that if a representative sample of the population were subjected to 500 r of total body radiation within a 24-hour period, approximately 50 percent of these people would die. The effects of less radiation are shown in Figure 4. Remember that these figures are prepared on the basis of total body radiation within a short space of time. We should also note that it is quite difficult for a person to receive a high radiation exposure unless there is gross violation of simple safety precautions.

Continuous Exposures

So far we have discussed a single radiation exposure. This type of problem is similar to the ordinary injury. A man gets on a rickety ladder, the ladder breaks, and he falls. He can die, he can be injured and recover, possibly with some permanent disability, or he may be lucky and have no ill effects at all. In any case, the incident is finished, and there are no more effects on the body. There is, however, another possibility with radiation—repeated small exposures over an extended period of time. What are the effects of exposures on the individual? What sort of radiation safety levels must we have so that there will be no apparent effect? What is the radiation level to which people can be safely exposed?

There are two factors which we must take into account. The first is the effect of the radiation on the individual. What damage is done to the individual by repeated small dosages of radiation over a period which might extend from the time he starts work until he retires?

It appears that high dosages of radiation received over a few years time can have some effect on the life span of the individual. Studies of a group of radiologists, some of whom received as much as 1,000 r of x-ray exposure, show an average life span 5 years shorter than for other physicians. There is as yet no conclusive evidence that low dosages spread over a period of years have any life-shortening effect. On the other hand, there is no evidence to indicate that there is a level of radiation exposure below which we can say there is no life-shortening effect at all.

Genetic Effects of Radiation

The other problem deals with the population and eventually with society. This refers to the genetic effect—the damage done to the genetic life stream of the population by the exposure of large numbers of individuals to ionizing radiation. The genetic effect of radiation is one which confuses laymen. We should first dispose of some common misconceptions. There is no relationship between the genetic effect and sterility or impotence. Radiation doses so high as to be nearly fatal can bring about sterility, which is the inability to conceive children. Impotence is the inability to carry on sexual relations, and radiation has no effect on this.

Another misconception is that all handicaps present at birth are genetic. Only about half of these handicaps are genetic in origin. The others are caused by disease or other factors.

There is also a misconception that there is a direct relationship between an exposure to radiation and the conception and birth of a defective child in any specific instance. It is true that radiation striking genetic material which is used in the conception of the next generation may cause a mutant which may appear in a future generation. However, we cannot determine the cause of any specific mutant. The mutant may or may not be caused by radiation. Even if it is caused by radiation, we cannot know the source of the ray. On the other hand, no measurable amount of radiation from any source is so small that we can say positively it cannot have a genetic effect. The geneticist studies the population as a whole, not individuals. The problem is one of protecting the entire population by keeping all exposure to radiation down to the lowest practical limits.

The portion of the population of greatest interest to the geneticist is that one which has yet to make its substantial contribution to the next generation. He is, therefore, interested primarily in the radiation exposure of individuals from the time of conception to the age of 30. The effect of radiation damage on genetic material is different from the effect of radiation damage on ordinary cell material. When genetic material is damaged, a pattern is damaged. Once a pattern is damaged, it remains damaged. Thus, it is possible that genetic material damaged by a radiation exposure in a person's early childhood may have its effect on his children. However, we must also remember that the only radiation exposure which is genetically significant is that which strikes the reproductive organs. Radiation exposure of other organs or other parts of the body has no genetic effect whatsoever.

Complicating the situation are the facts that all of us receive background radiation from cosmic rays and naturally occurring radioactivity, that we are all exposed to a certain average level of radiation from medical and dental x-ray procedures, and that we all get a small increase in our background exposure from fallout from atomic weapons tests, no matter by whom they are conducted. Table I lists some of the common radiation exposures we meet in everyday life.

Table I. Common Radiation Exposures in Roentgens (r)

Man Made		Natural	
Source	*Exposure*	*Source*	*Exposure*
Fluorscope	5 to 400 r	Cosmic rays (sea level)	40 mr per year
Chest x-ray	10 mr to 1 r	Granite rock	100 mr per year
Dental x-ray	20 to 65 r	Ocean water	50 mr per year
Pregnancy x-ray	20 to 65 r	Average soil	30 to 80 mr per year
Barium studies (x-ray)	10 to 20 r per min	Radon in air	120 mr per year
Diagnostic studies			
for heart disease	140 r	$^{40}_{19}$K in the body	20 mr per year
for acne treatment	500 to 1,000 r per treatment	$^{226}_{88}$Ra (bone)	40 mr per year
for malignant tumor (local dose)	3,000 to 7,000 r	$^{40}_{19}$K from people in packed crowds	2 mr per year
Wrist watch	40 mr per year		
Luminous dials in airplanes	1.3 r per year	Low grade uranium ore	2.8 r per year
Uranium mine	5.6 r per year		
Fallout	100 mr per year		
Television sets 3 ft. away	0.04 mr per hour	$^{14}_{6}$C in the body	1 mr per year

Protection from External Radiation

Any laboratory using dangerous levels of radiation will have trained people to set the standards of radiation exposure and see that they are followed. The means of protection against radiation hazards will not change. The degree to which we apply the means of protection will vary depending upon the level of radiation exposure permitted.

The means of protection from external radiation exposure are a combination of three factors; time, distance, and shielding. Protection is provided by:

1. Controlling the length of time of exposure.

2. Controlling the distance between the person and the source of the radiation.

3. Placing an absorbing material between the person and the source of the radiation.

We must use all three means of protection.

Time

The effect of time on radiation exposure is easy to understand. If we are in an area where the radiation level from penetrating external radiation is 100 mr per hour, then in 1 hour we would get 100 millirems of exposure. If we stayed 2 hours, we would get 200 millirems; if we stayed 4 hours we would get 400 millirems, and if we stayed 8 hours we would get 800 millirems of exposure. The term *rem* comes from the initials of the words "roentgen equivalent man." Thus, one rem means that a man was exposed to a total dose of one roentgen of penetrating external radiation.

Time is used as a safety factor in working with radioactive substances.

For instance, before working in a high radiation area, the work to be done should be carefully preplanned outside the hazard area. Then the minimum time is spent in the radiation area to do the work.

Distance

Figure 5 shows the effect of distance on radiation levels from a small source of penetrating external radiation.

As the distance from the source increases, the radiation levels (roentgens per hour, r/hr) decrease very rapidly. At a distance of one foot, the source in Figure 5 would give a man 1000 rem if he stayed there for one hour. At a point two feet away, the radiation level is $1/2 \times 1/2 = 1/4$ of 1,000 r/hr or 250 r/hr; four feet away, $1/4 \times 1/4 = 1/16$ of 1,000 r/hr or 62 r/hr and so on. We see that a very little distance can go a long way in increasing our safety factor. Conversely, as we get closer to the radiation source, the radiation levels get higher. For instance, at 6 inches

(1/2 of a foot) the radiation level will rise to $2 \times 2 \times 1,000 = 4,000$ r/hr. This emphasizes one of the most important rules of radiation protection: *maintain the maximum possible distance between yourself and the source of the radiation.* Particularly avoid touching materials which emit large amounts of penetrating external radiation.

Shielding

Remember that radiation is stopped when it knocks an electron away from an atom. If we wish to stop most of the rays before they get to us, we can place a material which has many electrons in its constituent atoms between ourselves and the source of the radiation. The more electrons there are in the material, the more radiation will be stopped. Figure 6 shows a rough comparison between the atomic constitution of lead and water. Notice that lead has many more electrons in each atom than does water. Therefore, lead makes a better radiation shield than water.

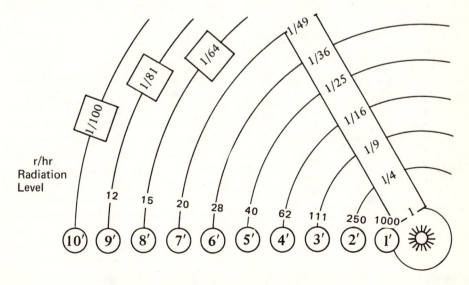

Figure 5. Effect of distance on radiation exposure

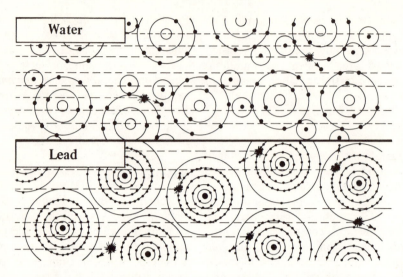

Figure 6.   Comparison of shielding capability of water and lead

Figure 7 shows the relative thickness of different kinds of shielding material that give the same protection. Lead, iron, concrete, and water are efficient in about the proportions shown on the chart for stopping the same amount of radiation. Different shielding materials are used in various applications. Lead, for instance, is quite compact and is most suitable where space is limited. Water can be used when it is necessary to see through the shielding material and to work through it with a long-handled tool to perform necessary operations

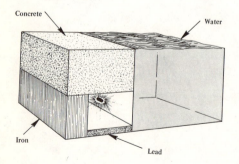

Figure 7.   Relative thicknesses required for equivalent shielding

such as sawing or cutting the radioactive materials.

Since we know that matter is mostly empty space, we realize that no material can be an absolute barrier to radiation. Regardless of the thickness of the material, we can see how some radiation can get through the material without hitting any electrons and, therefore, without being absorbed.

Half-Value Layer

In calculating the efficiency of shielding for external radiation exposure, we do not attempt to determine the amount of shielding necessary to stop all of the radiation, since no amount of shielding will do this. Rather, the measure which is used is that amount of shielding that it takes to stop half of the amount of radiation of a given intensity; this is called the **half-value layer** or the **half thickness.**

Examine Figure 8. At the left-hand side there are two sources of radiation, one giving off 25,600 mr/hr and the other giving off 204,800 mr/hr. The larger source has 8 times the strength of the smaller source. As we add successive half-value layers of shielding, the strength of the radiation emerging from the other side of the shielding is reduced to half of its entering strength. It takes 7 half-value layers of shielding to reach 200 mr/hr in the case of the source emitting 25,600 mr/hr.

We can see that there is always some radiation coming through the shielding. Shielding is designed to reduce the level of penetrating radiation to the worker. Therefore, we must always measure the amount of radiation coming through the shielding to be sure of safe working conditions.

Curie

The amount of radioactive material which is present is expressed in curies, millicuries, or microcuries. A curie is that amount of radioactive material which is disintegrating at the rate of 37 billion atoms per second. A millicurie is a thousandth of a curie, and a microcurie is a millionth of a curie. The curie bears no relationship to the weight of the material involved. If a material is very slightly radioactive, several thousand

pounds might be required to give one curie of radioactivity. Conversely, if a material is very highly radioactive, a fraction of an ounce of the material might produce a curie. For instance, one curie of cobalt-60 would weigh approximately 880 micrograms, while one curie of thorium-232 would weigh 10 tons.

The curie is not a measure of the radiation hazard from the material. It simply tells us the number of disintegrations taking place every second—not the kind of rays given off when these disintegrations occur. The problem is, "What happens when an atom of this particular material disintegrates?" The hazard depends upon the quantity of material and the type of radiation. A curie of one material might not give us any roentgens because it does not give off penetrating external radiation. So we can not measure the hazard until we know the number of curies involved and the identity of the material.

Short Range External Radiation

We have been discussing long range penetrating external radiation. There is another type of radiation hazard which

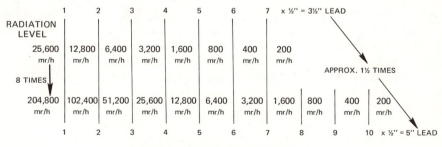

Figure 8. Typical effect of adding successive half-value layers of shielding

is less penetrating and has a shorter range. This type of radiation is an external hazard only when we come in extremely close contact with the radioactive materials either by handling them or allowing the material to be deposited on our bodies.

In Figure 9, we cannot tell which one of the 45 caliber automatics is loaded and which is not—both look alike. Some radioactive materials are quite safe to handle with bare hands while others are not. It is impossible to tell simply by looking at the material whether or not it is safe to handle with bare hands. The general rule is very simple: *Don't handle radioactive materials with your hands unless you know that it is safe to do so.*

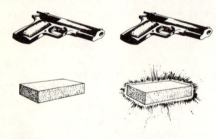

Figure 9.   Don't handle anything unless you know it's safe

Misconceptions about Radioactivity

Some people assume that radiation from radioactive materials will make other things radioactive. Think of the structure of the atom and the effect of the radiation from radioactive material. The radiation strikes electrons knocking them out of orbit. To make something radioactive, we must penetrate the nucleus of the atom because radioactivity originates in the nucleus. The rays from radioactive materials cannot penetrate the nucleus and therefore cannot make something radioactive.

Nomenclature

Thus far we have been discussing radiation without regard to the nature of the rays. The most penetrating type of radiation is called **gamma (γ) radiation.** This radiation is related to x-rays and ordinary light. Gamma rays have no mass but are very energetic. The less penetrating type of radiation is called **beta (β) radiation.** Beta rays (or beta particles) are high-speed electrons that have a range of only a few feet in air. **Alpha (α) radiation** is the nonpenetrating radiation. The alpha particle is a part of the nucleus of the atom that disintegrates. It cannot penetrate the outer layer of the skin and represents no external hazard.

Internal Radiation

The internal radiation exposure problem (Figure 1) is much more complicated than the external radiation problem. There are four possible ways to get radioactive materials into the body:
1. By breathing.
2. By swallowing.
3. Through breaks in the skin.
4. By absorption through the skin.

When a radioactive material gets into the body, the first question is, "how long does it stay?" A high percentage of anything we inhale is immediately exhaled. Materials which we swallow that are not soluble in the body's digestive system are discharged rapidly through the feces. If a material is soluble and is breathed into the body, it will go to the blood stream and is carried around the body to the various organs.

If an organ rejects the substance chemically, the blood stream will take it to another organ. If no organ will accept the material, the blood takes it to the kidneys, where it is disposed of through the urinary system. If the organ has a use for the material, or if the organ thinks the substance is similar to something it normally uses, it accepts the substance. For instance, the bones need calcium, and radium is chemically similar to calcium. Therefore, when the blood takes radium to those areas where new bone tissue is developing, the bone accepts the radium. Thus, the radioactive material is deposited in the bone.

Some chemical substances such as sodium and potassium are widely used throughout the body. If a radioactive form of one of these elements is introduced into the body, it is dispersed throughout the entire body. Other elements go to specific organs. For example, iodine goes to the thyroid gland. Body organs react to a substance as a result of its chemical nature without regard to whether or not the material is radioactive. All isotopes of an element react chemically the same way.

Half-life

The radioactive *half-life* of the material is important. The half-life is that amount of time that it takes half of the material to lose its radioactivity. If a material has a radioactive half-life measured in fractions of a minute, it will decay very rapidly. If the material has a long half-life (*e.g.,* thousands of years), then the decay rate is very slow.

The *biological half-life* of a material is that period of time which it takes for half of the material to leave the body. Some materials leave the body quite rapidly and will not do much harm. When we combine the radiological half-life with the biological half-life, we have the *effective half-life* of the material in the body. Standards have been established for the permissible burden of various radioactive isotopes in the body.

Protection from Internal Radiation Hazards

Each plant or laboratory that uses radioactive materials will have its own ways of protecting people from radiation hazards since each location has different problems. However, all work with radioactive materials must be done in special, well marked areas. We will mention some common ways to protect yourself from getting radioactive material inside your body (Figures 10 and 11).

Sometimes radioactive substances become airborne. Buildings are designed to make the air flow from the "clean" areas to the "hot" areas and then through filters to the outside. The offices, halls, lunch rooms, and restrooms are isolated from airborne radioactive substances. In areas with high concentrations of airborne radioactive substances, respirators or full face masks with independent air supplies are used. Protective clothing is worn over the rest of the body. Large amounts of radioactive substances are handled in special boxes (called glove boxes) to keep the radioactive materials inside.

However, the radioactive material is usually a solid or liquid. Rubber gloves are used to protect the hands and protective clothing is worn. Again, special boxes are used to contain the material.

Always remember that you can expose others to internal radiation. Be careful to keep radioactive materials in their proper place.

Contamination

Radioactive material which is spilled or gets out of the special working area can be a hazard. An alpha-emitter deposited on a wall or floor would present no radiation problem because of the short range of alpha radiation. How-

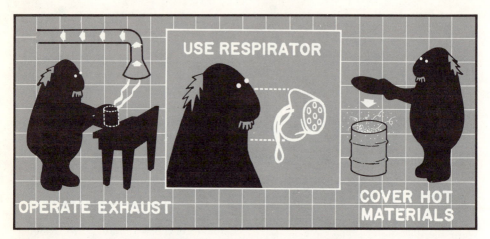

Figure 10.   Avoiding internal radiation

Figure 11.   Cleanliness avoids internal radiation

ever, alpha-emitting contaminants on a surface must be removed because there is always the possibility that they will become airborne and be breathed by people in the area. Beta- and gamma-emitters present both the possibility of becoming airborne and of creating high radiation levels around the contaminated area.

There are many ways to prevent the spread of contamination—special boxes, covering containers, plastic bags, etc. Carefully follow procedures for controlling materials, covering containers, and enclosing processes. Clothing which is restricted to a contaminated area should not be worn when going to a clean area. People working with radioactive materials should not leave the contaminated area for a clean area without first checking at a hand and foot counter or other monitoring device to assure they are clean. A hand and foot counter is a device which measures the radioactive contamination on the hands and feet. If a person is contaminated, he should not leave the area until he is clean.

In areas where contamination may occur, special preparations can make it easy to decontaminate. Absorbent paper is spread on table tops and on laboratory benches. Special strippable films are used on wall surfaces so that the paint can be easily removed when it becomes contaminated. Air conditioning systems should be carefully designed to avoid the spread of contamination from one area to another.

Cleanup

When an area has been contaminated, there are a number of ways to clean it. If the radioactivity has a short half-life, it may be sufficient to leave the area until the radioactivity decays to a low level. Usually the area must be decontami-

nated, and the materials used in the cleanup must be put in special containers. If the material is an alpha-emitter, it can be fixed to the surface by painting since the alpha radiation has a short range and cannot penetrate the paint. Floor tile can be replaced in order to remove contamination. In general, contamination is removed by scrubbing the surface with detergents which are best suited to remove the particular contaminating material. It is impossible to neutralize radioactivity as we might think of neutralizing an acid spill. The only solution to eliminating unwanted radioactive substances is to remove the material physically or fill the room with concrete.

Contamination is most easily spread during an emergency such as an explosion or a fire. In a fire, radioactive materials can spread very easily by the air currents caused by the fire. They can also spread through the air conditioning system, or if they are spilled on the floor, they can be carried around on shoes. Personnel working in an area contaminated during an emergency must be extremely cautious not to make the situation more serious by transferring contamination to the clean areas with their clothing, tools, equipment, etc. Procedures for handling emergencies must be organized ahead of time so that the contamination can be removed as quickly and safely as possible.

Detection Instruments

Radiation can be detected by instruments designed for the purpose. To the average person, all such instruments are geiger counters. However, a *geiger counter* is only one type of several radiation detection instruments. It is a low-level instrument with a maximum reading of 40 or 50 mr per hour. Most geiger counters will register both gamma and beta radiation. By closing a shield on the counter, beta radiation can be screened out and only the gamma radiation will be counted. Beta radiation is found by taking the open shield reading (beta and gamma), and then subtracting the closed shield reading (gamma only).

Higher levels of beta and gamma radiation are read with an instrument called an *ionization chamber*. Ionization chambers for civil defense use are designed to read as high as 500 r per hour.

Since alpha radiation has such a very short range in air and little or no penetrating ability, it must be detected with special alpha measuring meters. The measuring area of the meter must be brought very close to the contaminated surface. When an area has been contaminated with a pure alpha-emitter, the entire surface must be scanned thoroughly, inch by inch, to detect the location of the contamination. Since the alpha radiation has a very limited range, the contamination level at one point is no indication of what the contamination level may be at a nearby point. When the surface is too irregular for the flat surface of the instrument to be brought close enough for measurements, a piece of cloth or tissue can be used to wipe a measured area of the surface and measurements can then be made on the piece of paper or cloth.

A similar technique can be used for beta- or gamma-contaminated surfaces to determine how much is sticking to the surface.

All radiation detectors measure the rate that radiation is being received by the instrument at the time the instrument is read. If the instrument is removed from the radiation field, the needle will return to the background counts arising from naturally occurring sources.

Dosimetry

For the protection of individuals, devices which will measure the accumulated radiation exposure are available. These come in two types—film badges and pencil dosimeters. Film badges consist of a piece of dental x-ray film in a special holder which is worn on the person. The x-ray film is sensitive to radiation. When it is developed, the film will be darkened in proportion to the amount of radiation received. The film badge will indicate the radiation which the wearer has received only if it is always worn in the radiation area. Film badges should be treated like personal property and not exchanged between developments. Film badges provide a permanent record of radiation exposure.

Film takes time to be developed, so a pencil dosimeter can be used to give an immediate measure of the radiation dose. A pencil dosimeter is a device which will record the radiation which is received, starting from zero, to the limit of the scale of the particular dosimeter. Some dosimeters can be read directly by holding them up to the light, others must be read in a special reading device. Dosimeters will give false high readings if they are dropped or damaged, so they are generally worn in pairs.

An individual should always wear the proper personal measuring devices when in a radiation area. The device must then be protected from radiation exposure when it is not being worn.

8

Electrical Hazards

Why is a chemical technician concerned with electrical hazards? The most important reason is that the complicated nervous system in the human body is very sensitive to small amounts of electrical current. Also, electrical sparks or overloads can cause fires, explosions, damaged equipment, or ruined experiments. Chemical technicians work with many electrical and electronic systems and must be able to use them safely.

Accidents caused by improperly controlled electricity usually occur without warning. Things happen so fast that damage is done before corrective measures can be taken. This means the hazards must be recognized and avoided before accidents happen.

Causes of Electrical Injury

High voltage is mistakenly thought to be the cause of great danger from electricity. It is not the voltage but the current and how long it flows that is important. For most people, low voltage does not command the same respect as high voltage. However, there are hundreds of deaths every year from our low voltage, house lighting circuits—110 volts. In most laboratories you will be principally concerned with 110 volt circuits, but 220 volt circuits are common, and other voltage levels are occasionally

used. Safety fuses (and sometimes circuit breakers) are put into most circuits to interrupt the flow of electricity if the current exceeds a certain safe maximum. Safe maximum means a safe current for the equipment and normal wiring system, frequently between 15 and 30 amperes. But a 15 ampere shock to a human, regardless of voltage, is instantly fatal as we will see below. In fact, a fraction of one ampere may be fatal. Our familiar home and laboratory electrical circuits must be used with proper precautions.

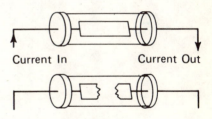

Current In   Current Out

Table I.   Common Electrical Hazards

Laboratory	Home
Lighting circuits	Lighting circuits
Ovens	Appliances
Heaters (room or equipment)	Power tools
Instruments	Radio, TV
Motor driven pumps, stirrers	Shavers, tooth brushes
Telephones	Telephones

Our exposure to electrical equipment is so frequent that we tend to forget the potential hazard. Table I provides a short list of familiar equipment to help you focus on the sources of electrical hazards. You should be able to add many more items.

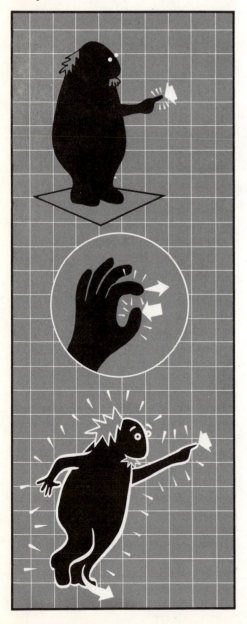

The Body as a Conductor

Receiving an electric shock means that the body or a portion of it becomes part of a conducting path for electrical current. The electricity enters the body from a *live* source, like a wire, and leaves some other part of the body through another conductor or conducting surface such as another wire, water pipe, or wet floor.

The electrical path through the body can be either local or general. The electrical path is referred to as local, for example, if current enters through one finger and exits through another part of the same hand. An example of a general electrical path might be that the current would enter one hand, go through the body and leave through the feet. The current path may be partly on the surface (skin) and partly internal. If a person's clothes are wet, some of the current may travel through the clothes. The electrical current will choose the path or combination of paths that offers the least electrical resistance.

What is meant by electrical resistance? A good conductor is said to have low electrical resistance. Metals like copper are good conductors. Dry wood, dry concrete, rubber, cloth, and plastics are examples of non-metals that are generally poor conductors or even non-conductors. Actually, not all rubber items are non-conductors of electricity. Some floor mats and shoes are made to conduct in order to reduce a build-up of static electricity.) Non-conductors have a very high resistance and are called insulators. If an insulator or poor conductor is wet with water, it may become a partial conductor. A perfectly dry wooden floor acts as an insulator. If the same wooden floor becomes wet, it may act as a conductor. By wearing dry shoes and standing on a dry wooden floor, you would probably feel only a mild and uncomfortable shock if you touched a live wire with your hand or another part of your body. Relatively little electrical current would flow from hand to the feet because it could not go anywhere from your feet. If your shoes and the floor were wet, much more electrical current would flow through you—possibly enough to cause a shock that could kill you.

Electrical Definitions

How much electrical current must pass through your body to injure or kill you? To understand the answer properly you must first know the simple relationship of the three principal characteristics describing the flow of electricity. The strength or force tending to make electrons flow through a conductor is called the *electromotive force* (also called the *potential difference*) or more commonly the *voltage*. The difference in electromotive force between two points is measured in *volts*. This is different from the *amount* of electricity flowing through a conductor. The current refers to the quantity of electrons that passes through a conductor in a given time period. The term used to describe the amount of current is *amperes,* often abbreviated *amps* or *A*. We have already mentioned that a good conductor has a low electrical *resistance*. Poor conductors have high resistance and non-conductors have such a high resistance that no current can flow. The term used to measure the degree of resistance to the flow of electrons is *ohms*.

> *volt*—unit for the potential difference or electromotive force
> *ampere*—unit for the quantity of electrons/time
> *ohm*—unit for resistance to flow of current

As an analogy to electricity flow, visualize water flowing through a pipe from a water reservoir. The force making the water flow is caused by pressure (volts in electrical flow); the amount of water flowing would be given as quantity or gallons per unit time (amperes in electrical flow); the resistance to water flow is determined by the size of pipe, its length, and the roughness of its inner surface (ohms in electrical flow).

The relation between force, current, and resistance is given by **Ohm's Law:**

$$\text{current} = \frac{\text{force}}{\text{resistance}}$$
$$\text{or}$$
$$\text{amperes} = \frac{\text{volts}}{\text{ohms}}$$

For a conductor with constant resistance (measured in ohms), the current (amperage) increases as the electromotive force (voltage) increases. The most important consideration for us is that *for a given voltage, the lower the resistance, the greater the current.* If the human body accidentally becomes a conducting path for electricity, any reduction in its resistance will allow a greater current to flow through the body if the voltage difference just stays the same.

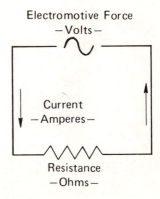

Electromotive Force
— Volts —

Current
— Amperes —

Resistance
— Ohms —

Electrical Resistance of the Body

What reduces your body's resistance? Wet hands, wet shoes, and a larger area of contact with the electrical conducting source will all reduce resistance. In many situations rubber or dry asbestos gloves will prevent a serious electrical shock when hands accidentally touch a live wire. Insulation at the point where the electricity will leave the body increases the total resistance to flow and reduces the total current flowing.

Let us look at the results of one experiment which illustrates the wide variation in resistance to current shown by a human touching two electrical sources, one with each hand:

a) one dry finger of each and on the electrode—resistance = 100,000 ohms

b) one moist finger from each hand on the electrode—resistance = 40,000 ohms

c) same as b) except fingers wet with salt solution—resistance = 16,000 ohms

d) tight grip, normal hands, one on each electrode—resistance = 1,200 ohms

e) same as d) except immersed in salt solution—resistance = 700 ohms

Effects of Electrical Shock

The effects of electrical shock summarized in Table II, are startling. These figures are approximations and vary with individuals and circumstances. Common sense dictates avoiding electrical shock completely. Current levels well below those listed can be very hazardous.

The victim of an electrical shock may lose control of practically all muscles. Any person caught in this situation will need instantaneous help. However, it is important that the right things are done by the person trying to help or he may also be electrocuted. The victim must be disconnected from the source of shock. If the source is a live wire it can be knocked or dragged aside with a dry board or other suitable, *well-insulated* tool. The victim must be rolled away

Table II. Degrees of Electrical Shock (1)

Effect of Electrical Shocks	Current, mA[a] 60 cps, ac	Current, mA[a] dc
Threshold of perception through skin	1	5
Threshold of muscular decontrol, let-go current[b]	6-9	70
Threshold of danger to life from failure of heart and respiration	25	80
Threshold of fibrillation	100	100

[a] mA = milliampere = 0.001 ampere. 60 cps ac = 60 cycles/sec alternating current (now called 60 Hertz), same as home lighting circuits; dc = direct current.

[b] As the current increases above the detectable level of 5 milliamperes (dc) it is more strongly felt. The human muscular and nervous systems react immediately to electrical shock. At very low current the reaction is to jump or pull away from the source of the shock. A special case involves grasping a live conductor. A shock through that hand tends to contract the muscles and as the shocking current approaches 70 mA (dc), control is totally lost. This is called the let-go current. It means that if one is grasping a live wire and receives that level of shock, he is unable to unclench his fingers and get away from the wire.

from the current source if it cannot be removed. If the rescuer touches the victim or the electrical source before breaking the circuit, he too will be electrocuted. Every year many well-intentioned rescuers become second victims by neglecting to break the circuit first. If possible, the electricity can be shut off by safely opening the right switch. Once freed, the unconscious victim needs immediate artificial respiration (*see* Chapter 1). While applying artificial respiration, attract attention and summon medical help.

At current levels near or somewhat above the let-go current a prolonged exposure is much more serious than a brief momentary shock.

Direct current (dc) and alternating current (ac) affect the body differently. Alternating current is more hazardous than direct current at less than 100 mA. However, a direct current arc will last a little longer and cause worse burns.

Circuits conducting alternating current with a frequency different from the common 60 cycles are sometimes used. Usually these currents are very high frequency and may go to millions of cycles

per second. In general the very high frequency circuits tend to cause rapid heating but not the severity of shock we discussed above. However, it is far safer to treat all electrical circuits as potential sources of severe injury.

Table II shows that at 25 mA (ac) or 80 mA (dc) breathing becomes difficult or impossible. This causes asphyxiation and results in partial paralysis of lungs and diaphragm. Immediate and continued artificial respiration (*see* Chapter 1) can save the victim who otherwise might not recover. At 100 mA the heart undergoes ventricular fibrillation (a complete upset to the rhythm of the heart beat), with uncoordinated fluttering and spasms. This is nearly always fatal.

Grounding of Equipment

Many modern electrical appliances are *grounded*. (It would be safer if all were.) This means that the case or metal housing is electrically connected to a special conductor which carries electricity away harmlessly to the ground. Water pipes are frequently used for grounding although regular grounding wire may also be available. Before using a water pipe for a grounding connection make certain that it does provide an ample circuit to ground. Some new installations utilize plastic dielectric couplings to reduce electrolysis and therefore do not provide ground. If some insulation wore off a wire in a hot plate and allowed the bare wire to touch the metal case, the case would become a source of a severe shock for anyone who touched it. However, the case can be grounded so that the stray current will be carried away to ground instead.

You will find a three-pronged plug on many instruments and appliances today. Two of the prongs carry the current and the third makes a ground connection to the case. The outlet into which the plug is inserted must have the corresponding three holes. It is very unsafe to bend or mutilate the three prong plug to fit a two hole outlet. Adapters are available which allow for grounding the third prong independently from the two hole outlet. Be sure that the ground wire of the adapter is securely grounded to the screw in the receptacle (*see* photograph).

Laboratory electrical equipment is usually specially manufactured for safety in a chemical laboratory environment. Switches and motors are often explosion proof to prevent sparks. The line cords

should be the heavy duty three-wire type with three prong plugs to assure good grounding.

Everyone who uses electrical equipment should know about the Underwriters' Laboratories, Inc. It is a nonprofit corporation organized to inspect and test representative samples of materials and devices. These devices are labeled as safe for their intended use or rejected as hazards to safety. Although not all equipment is submitted for testing, those pieces that have been inspected and use the UL label can be depended upon to be safe as long as they are not abused or worn to the point of hazard. Note UL labels carefully. A label on a plug or cord may mean only that the plug or cord has been tested. A label should also be on the instrument or appliance.

Summary

The most important points discussed above can be briefly summarized as follows.

Severity of electrical shock is determined by:

1. Amount of current flowing through the body

2. Path of current through the body

3. Length of time the circuit is maintained

4. Type of electrical circuit, dc, ac (frequency)

Amount of current is determined by:

1. Voltage of the circuit

2. Insulating qualities of the place in which the individual is located at the instant of contact

3. Resistance of skin or clothing

4. Area of contact with the live conductor

5. The pressure of contact with the live conductor

Safety Precautions with Electricity

1. Equipment should be in working order, and only properly designed equipment should be used.

2. In preparing an electrical device for use in the laboratory, check these points:

a) Inspect the unit to be sure wires and other electrical components are not frayed, loose, or broken. Faulty insulation on appliances must be repaired or replaced.

b) Make sure the bench or area on which the unit will be installed is dry.

c) If the unit is a heating device, make sure the area is free of flammable liquids, vapors, or open containers of flammable liquids.

d) Turn power switches on the unit to **OFF** before plugging the unit into an outlet. This prevents sparking at the plug.

e) Don't handle any electrical device when your hands are wet.

3. Provision for proper grounding must be made and used.

4. If you have to work on any electrical circuit or repair faulty insulation, *disconnect* the appliance instrument. Don't just turn off its switch. If this is not possible, open the circuit main switch and secure it in the open position or tie a note to it to prevent someone else from closing it while you are exposed. Don't guess about whether a circuit is alive or dead. Test it with proper test equipment.

5. Use proper safety equipment when necessary: rubber gloves, rubber mats, insulated tools, etc.

6. Replace a burned-out fuse with a proper fuse. If it burns out find the cause. **Never** substitute a conductor (a nail or penny) for a fuse. Do not use an oversized fuse.

7. Electrical work should normally be done by qualified electricians familiar with both the requirements of the job and the electrical codes.

8. Equipment must be inspected periodically.

9. If an extension cord must be used, make certain that it is heavy enough to carry the power safely.

10. Do not store containers of flammable volatile liquids in refrigerators which have not been made explosion proof.

11. If electrical equipment must be left unattended, inspect it before leaving to make sure it is working correctly and reliably.

12. Be thoughtful and careful in the day-by-day handling of electricity and appliances; *i.e.,* don't jerk plugs from wall or bench outlets by pulling on the cord; grasp the body of the plug and pull straight out; don't use equipment for jobs for which it was not designed; stay constantly alert to any malfunction and correct it immediately.

Literature Cited

1. Pieters and Creyghton, "Safety in the Chemical Laboratory," Butterworth, London, 1957.

TREAT ALL ELECTRICITY
WITH RESPECT

9

Compressed Gases

Gases are normally stored in heavy metal cylinders of various sizes. These cylinders are labeled and color coded to identify the contents. The gas is released through a high pressure valve which is protected by a domed cover when the cylinder is not in use. A few typical gas cylinders are shown in Figure 1. The label on the cylinder will state the name of the gas it contains, the grade or purity of the gas, the pressure to which it has been filled, and safety data (flammable or corrosive). Figure 2 shows two such labels. The color codes may differ and should not be used as the final means of identifying cylinder contents.

With some exceptions, gas cylinders are filled to very high pressures (1000–

Figure 1. Cylinders of compressed gases

2000 psi). The weight of large cylinders, plus the dangers of releasing high pressure gas means that these cylinders must be handled very carefully even if the gas inside is as harmless as helium or air.

Transporting Gas Cylinders

If you refer to data in a gas products catalog, you will see that size 1 cylinders weigh about 200 lb and size 2 cylinders weigh about 100 lb. Whenever you move these larger cylinders, always use a strong wheeled cart especially designed for this purpose. The cylinder must be securely fastened to the cart during transport as shown in Figure 3. Cylinders are sometimes mounted in roller-equipped base rings which can be used for moving cylinders short distances, but this is not considered good practice.

Before moving the cylinder, be sure the protective dome is properly fitted on the tank. If a regulator is attached, remove it and replace the dome.

When you reach your destination with a size 1 or 2 cylinder, tilt it to an upright position, remove it from the transport cart, and *immediately* fasten the cylinder to a wall or work bench by straps or chain devices such as those shown in Figure 4. *Never work with a cylinder that is not fastened or chained.* The

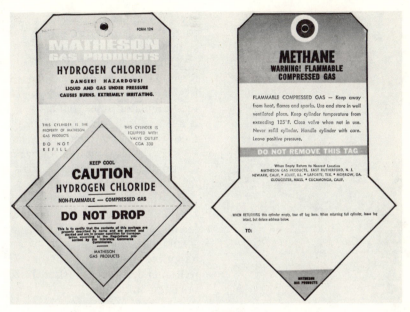

Figure 2.   Labels for compressed gas cylinders

fasteners should be placed at a point more than halfway up the cylinder but not at valve level.

Smaller cylinders should also be transported with safety domes in place. They may be hand carried (a size 3 cylinder weighs about 40 lb) but should be installed in a rack or holder and securely fastened.

Gas Cylinder Regulators

The tank regulator is a device for controlling the flow of gas from a cylinder. High pressure gas cylinders should never be used without a proper regulator. A typical regulator as shown in Figure 5 consists of:

a) a connection for attachment to the cylinder

b) a high pressure gage (0–3000 psig) to indicate the pressure of gas in the cylinder.

c) an adjusting screw or valve to regulate the output delivery pressure

d) a low pressure gage (0–100 psig) to indicate the output delivery pressure

e) a needle valve to give fine control on the flow of the gas

f) a hose connector for attaching rubber or metal tubing.

Proper selection of regulators is very important and sometimes very critical if accidents are to be avoided. Consideration must be given to the following rules. Failure to observe proper precautions could result in a serious accident.

1. Be sure the high pressure gage on the regulator has a capacity greater than the cylinder pressure. If you do not know the normal tank pressure, refer to the supplier's catalog or contact the supplier.

2. The regulator should be capable of reducing the high pressure to the desired operating level. The low pressure gage should have a suitable scale. For example, if you must read 3 or 4 psig with reasonable accuracy, then a 100 psig gage will not be satisfactory. How-

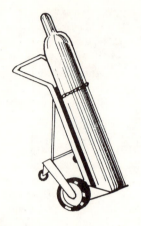

*Figure 3.   Cart for transporting gas
cylinders*

Figure 4.   Gas cylinder fastener

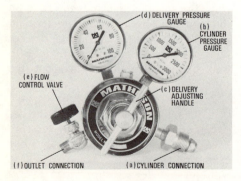

Figure 5.   Gas cylinder regulator

ever, if you need a pressure of 25 or 30 psig, then the 100 psig gage would be suitable.

3. You will find that cylinders of different gases have different types of outlets and different screw sizes. The correct regulator must be chosen or the connector will not fit. Unfortunately, there is no uniform code between different suppliers so that a nitrogen regulator for one supplier's cylinders may not fit those of another supplier.

4. Adapters can be obtained to fit a regulator onto a tank which has a different outlet. In general, this practice should be avoided since it bypasses an important safety precaution. Only a well-qualified expert should consider breaking this rule.

5. *Never use a regulator on an oxygen cylinder that has been used for some other gas. This is a most important rule that must not be violated. The high pressure oxygen will react with any oil in the regulator and this reaction may lead to a serious explosion or fire.*

6. Regulators for corrosive gases such as hydrogen chloride must be made from inert materials. Stainless steel and Teflon coated metals are common examples of special inert materials. Before using corrosive compressed gases, check with your supervisor for a standard set of precautions and recommendations.

7. Cylinders of flammable gases should be grounded to eliminate the possibility of a spark's igniting the gas.

Using the Gas Cylinder Regulator

When the gas cylinder is securely strapped in place, remove the dome protector and clean any rust or dirt from the threaded cylinder outlet. If the outlet is covered by a dust cap, remove the cap. Connect the regulator to the tank and tighten the coupling with a large crescent wrench.

Before opening the high pressure valve, be sure that the adjusting valve for regulating the output pressure (*see* Figure 5) has been turned counter-clockwise until there is no resistance to turning and the downstream gage reads zero. The needle valve should be closed.

Now open the high pressure tank valve by turning it counterclockwise. *Never stand directly in front of the regulator when opening the high pressure valve.* When the valve is opened, the gas may surge (sudden movement that may be damaging) into the meter. If there is an explosion or fire, it will come directly from the front of the regulator. If all precautions have been followed, there is little danger. However, it is still wise to stand to the side when opening the high pressure valve.

Turn the main cylinder valve fully counterclockwise and then back one-half turn clockwise. This is a practice which lets you safely and easily determine if the valve is in an on or off position. If the valve is tight, it is closed; if the valve is loose, it is open.

When the tank valve was opened, the high pressure gage (*see* Figure 5) indicated the tank pressure. You can now turn the regulator valve clockwise until the desired delivery pressure is indicated by the low pressure gage.

Gas can now be delivered into the chemical system or gas vessel by attaching the system to the hose nipple or metal fitting. The needle valve is used to control the flow of gas conveniently.

To shut off the regulator use the following procedure:

1. Close high pressure valve. (Turn clockwise until tight.)

2. Open the needle valve so that the gas in the regulator flows out. Both gages will return to zero.

3) Open the regulator valve. (Turn counterclockwise until the valve is loose.)

4. Close the needle valve.

A high pressure system should always be shut down when it is no longer in use, *especially if the gas is corrosive. This is another cardinal rule that must not be violated.*

To remove the regulator, the cylinder should be shut off as described above. Remove the regulator with a crescent wrench, replace the dust cap, and replace the safety dome. Do not let cylinder pressures run down to zero before replacing the cylinder. Mark empty cylinders prominently, "M.T." By returning the cylinder with some pressure, the gas supplier is assured that no impurities have been introduced and he can safely repressurize the tank with the same gas at a high purity level.

Prevent leaks. If a leak is suspected, check at once with a soap solution or other safe detecting means. To detect leaks, a soapy water solution is brushed onto the tank fittings, the regulator, and the attached tubing. A leak will cause the formation of bubbles which will grow in size as gas escapes. If a leak is found in a corrosive gas system, remove the cylinder *safely and quickly* to a safe place—preferably out-of-doors—and inform your supervisor promptly.

Storing Compressed Gases

Always put gas cylinders in storage areas designated for the particular gases being used.

Cylinders containing hydrocarbon bases are always stored out-of-doors.

Do not lay cylinders on their sides unless data from the manufacturer specifically indicates that the cylinder design will allow withdrawal of gas safely in this position. Acetylene fuel cylinders used on atomic absorption apparatus present a special hazard. Acetone is used as a stabilizer in the cylinder and can

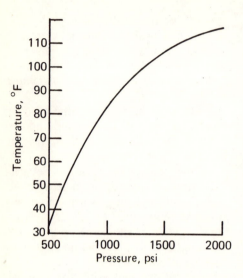

Figure 6. Temperature vs. *pressure for carbon dioxide*

be withdrawn with the gas causing serious fires in the apparatus.

Never store cylinders in bright sunlight or near heat sources. 125°F is the highest safe storage temperature. If cylinders get hot, the internal pressure may rise to a dangerous level. Cylinders are provided with safety devices, but the release of a large volume of any compressed gas is a hazard even under non-explosive conditions. The change of pressure with temperature for carbon dioxide is shown in Figure 6. In the case of carbon dioxide, about 75% of the gas in the cylinder is compressed to a liquid. This allows the supplier to fill the tank with almost twice as much carbon dioxide, but if the temperature increases too much, the liquid is converted to gas, and the pressure will exceed the safety point.

Gas Cylinder Safety Devices

Compressed gas cylinders are protected by safety devices placed near the main valve. Cylinders containing hydrocarbon gases (propane, butane, etc.) have a small spring-loaded safety valve set to open at several hundred pounds per square inch pressure. Some gases, such as chlorine, hydrogen chloride, and acetylene, have one or two fusible plugs designed to melt at 165°F or 212°F. In case such a cylinder is exposed to fire or high temperature, the fusible element will melt and allow the gas to escape. This reduces the internal pressure and protects the tank from explosive rupture. Certain flammable gases such as hydrogen are handled in cylinders fitted with a frangible disc designed to burst and allow the gas to escape if the pressure gets beyond some established safe limit. Depending upon the gas, this limit may be as high as 6000 psi. Compressed free gases, such as nitrogen, oxygen, helium, and argon, are fitted with a similar small, frangible disc, sometimes in combination with a fusible plug.

Neither of these safety devices should ever be altered in any way. If there is any reason to suspect that they are not in good working order, do not use the cylinder. Report your suspicion to your supervisor.

Gas Cylinder Fires

Should a fire start in a system which is fed by a cylinder of compressed flammable gas, shut the tank valve before putting out the fire. Otherwise the tank will continue to release the flammable gas which can then ignite or explode.

Data sheets are available on all compound gases and will indicate the specific handling problems encountered for specific gases. Data sheets or the manufacturer's representative should be consulted before attempting to handle any compressed gas with which the user is unfamiliar.

10

Laboratory Notebooks

Can you remember some experiment or test you did a month or so ago? Can you recall exactly what you did and observed? What was the purpose of the experiment? What were the names of your co-workers? Exactly what procedure did you use? Did everything go smoothly? What were your results? How did the results correspond to the reasons for performing the experiment? Are you sure you didn't make some errors in your work which you would like to recheck? Did your results (or lack of them) suggest any further experiments?

Could you repeat the experiment exactly or repeat it and change just one or two parts? If you were required to describe the experiment precisely in a letter, could you do so? If someone had reasons to repeat your work exactly, using only your written description, could he do it? Finally, if a few years from now, some other person wanted to know exactly what you did, could he tell from your written record only? These questions are very important to a chemical technician and can be answered affirmatively only if the information is recorded in a proper notebook.

Student Notebooks

A student's notebook can be relatively simple and still be adequate. A chemical technician must keep a complete and accurate record of his work in much more precise form. It makes sense for a student to keep his notebook records somewhat similar to those required in actual laboratory practice. The necessary professional standards will be easier to follow later when they are required.

Use a bound book, not loose leaf pages. The first page should show your name, the chemistry course name and number, and the date on which the course was started. This should be followed on the next page by a "Table of Contents," which will list each experiment and the page on which it begins. Then each individual experiment should be described separately. Use the following form when it applies:

1. Page number
2. Title of experiment
3. Object of experiment
4. Equations of reactions (if any)
5. Table listing principal reactants and products, their formulas, and important physical properties
6. Procedure followed
7. Data and observations
8. Calculations (if any)
9. Conclusions (and interpretations where they are appropriate)

In using the recommended form, there are several conventions that should be observed. Only write on the right-hand page. Blank lines or unused portions of a page at the end of an experiment should have a diagonal line crossed through them to show that nothing was added at a later time. At the end of each day's entry, add the date and your signature. Some employers require initials, and they keep a list of initials for identification.

Start the practice of entering data and observations immediately. Do not write information on odd pieces of paper to copy into your notebook later. Keep your writing legible, accurate, neat, and clean. Use ink so that the record is permanent. Ball point pens are better than fountain pens because the ink will not run if liquid is spilled on the notebook. Never erase. Instead, draw a line through unwanted entries so they can still be read. You may want to think about the bad results as well as the good ones. Lined-out results can be read—erased results are lost forever. When possible, prepare tables ahead of time to accept the data and observations so that the readings or comments can be entered simply and quickly. If the notebook entries for one experiment are to be continued on some page further along in the book and the next page has already been wholly or partially used, indicate which page to turn to and, at that new page, indicate the number of the page from which it is continued.

How detailed should the entries be? Include everything that might be necessary for another person to duplicate your experiment. Try to be brief but include everything someone else might need to know. On the other hand, omit details that he would know in any event or those for which his choice might be just as good as yours. (This is a tricky point, though. Later work may show certain entries to be very important even though initially they do not seem to be.) No

supervisor has time to read long-winded procedures or conclusions that could have been condensed easily with no loss of important information.

Finally, enter data and observations exactly as you read or observe them. Never fudge figures or change data because you think that is the way it should have been. If an experiment is ended by an accidental spill or breakage, say so and start it over again. The essence of a notebook is accuracy. It is a permanent record of what you did and what happened.

The Industrial Notebook

Nearly all of what has been said above concerning the form of a student's notebook applies to the industrial notebook. On the job there are additional requirements for reasons that will be explained later.

Many employers have forms and instructions concerning notebooks that are kept by employees. These forms and instructions must be studied and followed precisely. There are some differences about how notebooks are kept among employers and even between laboratories, plants, and related chemical facilities of the same employer. These differences usually introduce only minor variations on the guidelines that follow. In addition to the form suggested for the student notebook you will have to:

1. Enter project numbers on each page.

2. List the names of all participants directly connected with the experiment, including your own name and your supervisor's name.

3. Enter each experiment in chronological order in the Table of Contents.

4. Make cross-references to related material in the same or other notebooks.

5. Explain all lined-out errors and corrections in the margin. Date the er-

200

| SUBJECT NO. | 127-62 | CROSS REFERENCE | CLR 6403/69 | | p. 300 |

PARTICIPANTS *George F. Brown and John A. Doe*

OBJECT OF WORK *Determination of the effectiveness of various salts as catalysts for curing epoxy resins with a benzophenone tetracarboxylic dianhydride.*

Requested by George F. Brown per oral instructions on November 20, 1968 1969 J.A.D. 11/30/69 G.F.B. 11-30-69

Composition (Parts by weight):

Polyepoxide (EPON® RESIN 1005)	68.48
3,3',4,4'-benzophenone tetracarboxylic dianhydride	11.52
Filler (Silica 219, silica sand)	20.00
Catalyst (varied)	1.0 *

* 1 part per 100 parts by weight of polyepoxide

PROCEDURE

Compositions were prepared by dry blending the above components in a ball mill until the composition was pulverized to 50 100 mesh or finer

11/30/69 J.A.D. 11/30/69 G.F.B.

These compositions were then evaluated for impact resistance, solvent resistance, and flexibility.

A fluidized bed was prepared from the above compositions and a coating having a film thickness of 10 mils was applied to 3/4" diameter, 6" long sandblasted round steel probes. The coatings were then cured for 30 seconds at 450°F.

TEST METHODS

Impact Resistance - determined on the cured films by direct impact measured in inch-pounds required to shatter the coating.

Solvent Resistance - determined by immersing the coated article in methyl ethyl ketone (MEK). The time was then noted when the

— John A. Doe - 11/29/69 —

ALL ENTRIES MUST BE SIGNED AND DATED DAILY   LR Nº 7083

Figure 1.   A sample industrial notebook, page 300

SUBJECT NO. 127-62	CROSS REFERENCE CLR 6403/69	p. 301

PARTICIPANTS George F. Brown and John A. Doe

OBJECT OF WORK (Continued from page 300)

surface coating could be scratched with the fingernail (minutes).

Flexibility - A No. 10 copper wire was coated with the above compositions to a film thickness of 10 mils. The cured wire coating was then clamped firmly on a 1" round mandrel. The free end was then bent around the mandrel until the first crack appeared on the wire coating. The angle at which such cracking occurred was measured in degrees.

TEST RESULTS

CATALYST	IMPACT RESISTANCE (IN-LBS)	SOLVENT RESISTANCE MEK (MIN.)	FLEXIBILITY (DEGREES)
Stannous octoate	25	30	> 90
Lead octoate	10	5	65
Aluminum octoate	10	5	90
Calcium octoate	10	5	55
Manganese octoate	10	5	30
Cuprous octoate	10	5	30
Zinc octoate	10	10	75

John A. Doe 11/30/69

Read & understood George F. Brown 11/30/69

ALL ENTRIES MUST BE SIGNED AND DATED DAILY   LR N° 7083

Figure 2.   *A sample industrial notebook, page 301*

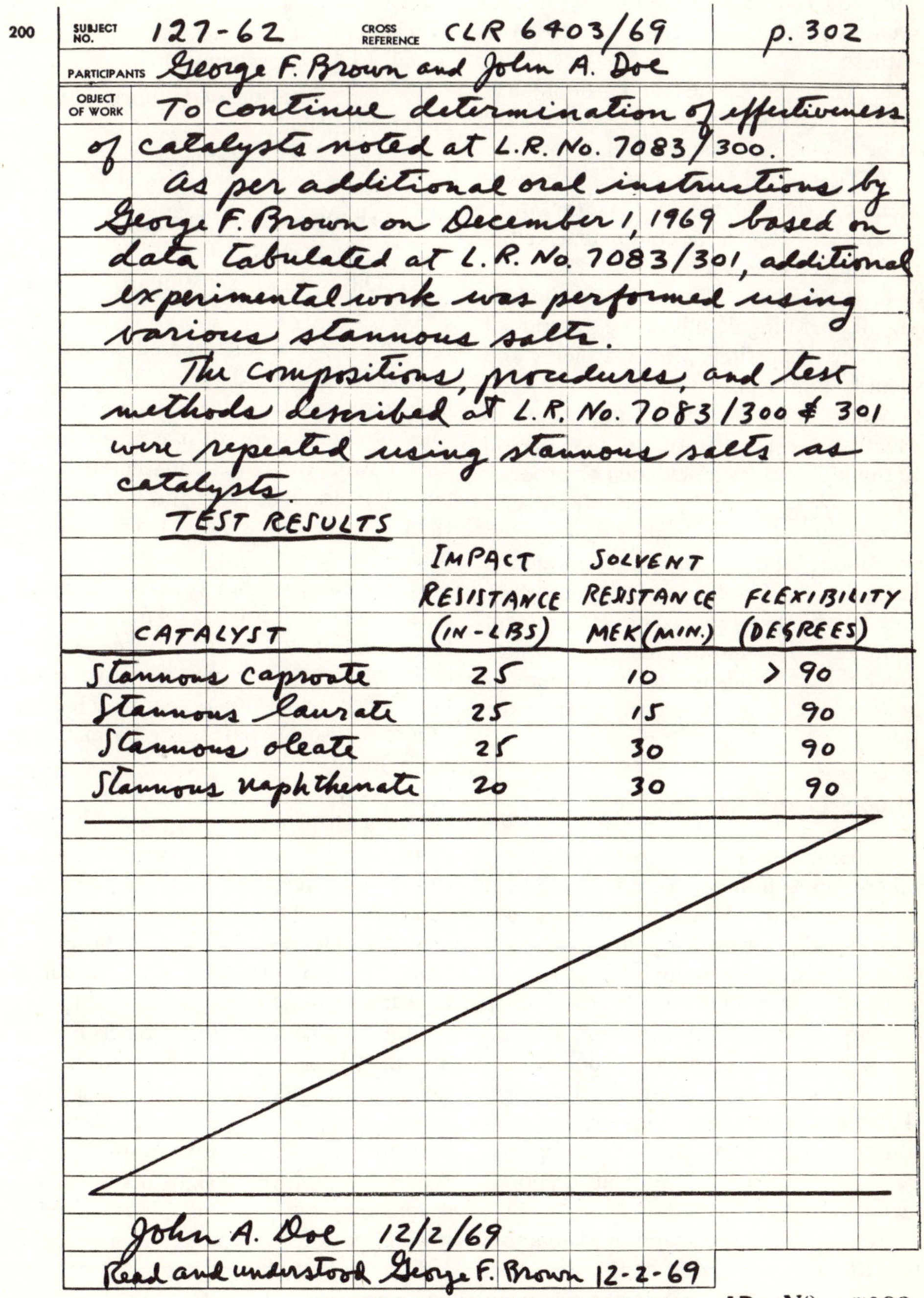

200

| SUBJECT NO. | 127-62 | CROSS REFERENCE | CLR 6403/69 | | p. 302 |

PARTICIPANTS George F. Brown and John A. Doe

OBJECT OF WORK To continue determination of effectiveness of catalysts noted at L.R. No. 7083/300.

As per additional oral instructions by George F. Brown on December 1, 1969 based on data tabulated at L.R. No. 7083/301, additional experimental work was performed using various stannous salts.

The compositions, procedures, and test methods described at L.R. No. 7083/300 & 301 were repeated using stannous salts as catalysts.

TEST RESULTS

CATALYST	IMPACT RESISTANCE (IN-LBS)	SOLVENT RESISTANCE MEK (MIN.)	FLEXIBILITY (DEGREES)
Stannous caproate	25	10	> 90
Stannous laurate	25	15	90
Stannous oleate	25	30	90
Stannous naphthenate	20	30	90

John A. Doe   12/2/69

Read and understood George F. Brown 12-2-69

ALL ENTRIES MUST BE SIGNED AND DATED DAILY   LR N⁰ 7083

Figure 3.   A sample industrial notebook, page 302

rors, add your initials, and have the corrections read, dated, and initialed by a witness.

6. Make all entries in chronological order. Do not leave spaces or blanks to be filled in at a later date.

7. Describe the object of each experiment clearly and briefly. As the work progresses, record a complete, factual account.

8. State procedures precisely, including all operating details and equations of reactions if possible. Record yields and identify products.

Many times it is very important to establish the identity of the first person to think of and try a new idea or process. In a college it might mean fame and recognition. In industrial or commercial operations and in some governmental laboratories, it is the basis for progress and completion reports on government contracts. The importance of complete notebooks should never be underestimated. They are an integral part of the job.

Patent Law

Laboratory notebooks serve functions other than as primary records in the field of patent law. However, if they are maintained properly for patent purposes, they will also serve many other purposes. The reverse is not necessarily true. Other primary sources of information include "conception of invention" notebooks, trip reports, and original drawings, sketches, flowsheets, etc. Secondary sources include letters, formal reports, purchase orders, analytical reports, memos, etc.

Almost without exception patentable inventions result from hard work and careful observation rather than from a flash of genius. Of the 18 or 20 U.S. inventions frequently listed as having

been of the most use to man, and on which very large industries were established, four were chemical:

1) Vulcanization of Rubber, Charles Goodyear, 1844;

2) Electrolytic Reduction of Aluminum, Charles N. Hall, 1889;

3) Thermosetting Plastics (Bakelite), Leo H. Baekeland, 1909;

4) Oil Cracking, William Burton, 1913.

A patent may be granted to anyone who is the first to discover or invent "a new and useful art, machine, manufacture, of composition of matter, or a new and useful improvement thereof." A patent, if granted, gives the inventor the right to prevent others from making, using, or selling the invention for 17 years. This does not deprive others of something previously theirs but describes something which never existed before and prevents piracy of inventive creations. The owner of a patent may license others to use his patent, or he may sell his rights.

If a company employs a technician to do research or other creative work, who owns a patent which may result from the work? This question has generated some controversy. The employee may be the inventor. However, an agreement at the time of employment usually specifies how the ownership of patent rights will be resolved unless the subject of the invention is clearly not related to the job assignment, and the invention was conceived and developed away from the job site on the employee's own time. When an employee assigns his invention to the employer, the employer applies for and pays for the patent application.

For further information, there are publications available from the Superintendent of Documents, U.S. Government Printing Office, Washington, D.C. 20402.

Three inexpensive pamphlets that are interesting and informative are *The Story of the United States Patent Office, General Information Concerning Patents,* and *Patents and Inventions: An Information Aid for Inventors.*

The National Association of Manufacturers has available upon request several pamphlets on the history, importance, and significance of patents.

There is no doubt that a chemical technician can be creative in his job. But creative effort must not be confused with invention. Performing some test or operation in an excellent manner is doing a fine job, but it is not invention. The process of invention must start with a completely new idea, but that is not enough. The device or process must operate as intended and must be described completely. The several references mentioned will tell whether the result of a thought and effort is appropriate for a patent application.

Are patents really important? They certainly are. A real example concerns two large companies, A and B. Both were subsidiaries of the same parent company and independently filed for a plating process patent. To avoid interference, Company A withdrew its application to allow B the rights, presumably with some understanding concerning later sharing of privileges. Company B was awarded the patent. Later the notebook records showed clearly that Company A's employee was really the inventor. Consequently, the patent was not valid as issued, neither A nor B got a patent, and Company A went out of business.

Why be so fussy about patents when such a large proportion of them result in no economic gain, often not repaying the cost of procuring them? Very often, the eventual value of a patent is obscure or unknown. When the Prime Minister of England, Mr. Gladstone, saw Faraday's device for converting mechanical energy into electrical energy by induction, he asked, "Will this ever be of use?" Faraday is alleged to have replied, "Some day you will collect taxes from it." Faraday had actually invented the first electrical generator.

Without saying more about the patent law, the final word to a chemical technician concerning his notebook should be to make certain that he understands and follows his supervisor's and employer's instructions and records data carefully and accurately.

Countless patent applications have failed, many excellent opportunities have been overlooked, severe accidents have been caused, and thousands of experiments have been repeated because of carelessly kept notebooks. How would you feel had you been the technician involved in the following true incident? He thought he had a poor analysis of the water content of a perchlorate oxidizer used for the propellant in a missile. He fudged the water content figure because he thought it was too low and said that he didn't want to appear to be dumb. It later resulted in a serious explosion and endless hours of tracking down the error.

11

Aids for Calculations

An important part of all technicians' work is calculating numerical answers. By the time he sits down to calculate an answer, much money, time, and effort have already been spent on a project. It is simply foolish to waste all experimental work by slipshod methods, careless errors, unproved shortcuts, poor reporting, or misplaced results.

Laboratory calculations do not require great mathematical skill. It is seldom necessary to derive equations using algebra, calculus, or more advanced operations of mathematics. Most laboratory calculations involve using a known relationship to combine known constants and experimental values to find an answer. Most of the time this involves only the four basic arithmetical operations: addition, subtraction, multiplication, and division. Occasionally, there is a logarithm or exponent to handle. There are also other aspects of making calculations in which good practices can help to avoid making errors.

Limits of Precision

Assume you are a technician working in a UV spectrometry lab and are asked to measure the purity of a new plant raw material, para-nitroaniline (p-NA). You are provided with a procedure from the manufacturer. You perform your measurements, and the calculation from your data gives a result of 101% for the p-NA assay. This worries you. You check and find that the chemical lab of your company has a technician analyzing the same material and he finds 98.7% p-NA. Now you are even more worried. Then you find out that the purchasing agent for your company has sent a copy of the vendor company's results which do not agree with either of the first two numbers. The laboratory supervisor summarizes all the results, shows them to you, and tells you that you did a fine job. How can this be when all four numbers were different?

Assay for para-Nitroaniline (p-NA)

	UV Lab	Chem Lab
Your company	101	98.7
Vendor company	98	98.8

Because of his background, training, and the method's precision statements, a lab supervisor knows that all test methods have limits. For example, the precision and accuracy of your UV method did not allow you to make a better measurement. On the basis of these considerations the summarized data really mean that the assay is about 98.7% p-NA.

The numbers are viewed in terms of ranges which are calculated from the precision statements in the methods used.

	UV Lab	Chem Lab
Your company	101 ± 3	98.7 ± 0.2
Vendor company	98 ± 3	98.8 ± 0.2

The lab supervisor and other managers of a company make judgments of this sort every day which affect all phases of the company's operations and its relationship to other firms. Judgments are also made about the performance of the technicians and other data producers. You can help them enormously in their jobs and be of much greater value to your employer if you understand the basis for such judgments. The following section explains how to use data to obtain meaningful results to experiments.

Rounding Off

If you were asked to count the number of triangles in the figure above, there would be no concern with approximations—there are five; not about four, or almost six; but five. Most measurements are not so simple, as was shown in our % *p*-NA example.

Sum of Impurities (%)

Water	0.123
o-NA	0.9
p-NCB	0.2
	1.273

p-NA Assay (%)

Total	100
Impurities	1.273
	98.727

Suppose the only major impurities in *p*-NA are water, *o*-NA (ortho-nitroana-line), and *p*-NCB (para-nitrochlorobenzene). The purity of the *p*-NA sample can be calculated by taking the sum of the % impurities and subtracting it from 100 to give a 98.727 assay. However, when this number is reported, it is rounded off to 98.7%. This is done because of the limits of precision in the data. Maintaining meaningful degrees of accuracy is the basis of the concept of significant figures.

Significant Figures

When obtaining data from an experiment, the last figure of each observation is usually an approximation. Because of the various techniques a laboratory worker could use to get all the needed data for a problem, there could be a great range of accuracy. One set of data could be reported to three decimal places while another method would only allow precision to one decimal place.

In the previous example, results for both *o*-NA and *p*-NCB have only one decimal place, or **significant figure** each (0.2,0.9) while the water analysis is expressed using three significant figures (0.123). The sum of the impurities and the final assay must contain only one decimal place because the least precise figure contains one decimal place.

In addition and subtraction, the result is rounded off to the least number of decimal places given in the data. The last two figures of the water impurity entry are useless because the other entries are precise only to one decimal place. The greater accuracy of the two extra decimal places in the answer is wasted and meaningless, so the answer is rounded off to one decimal place.

In multiplication and division, a product or quotient can have no greater percentage certainty than any factor in the computation. Suppose the GC (gas chromatography) method for *p*-NCB analysis is based upon the following calculation:

$$\%p\text{-NCB} = k\ h\ w$$

where:

k is a constant experimentally determined to be 0.231

h is the height in millimeters for the GC peak

w is the width in millimeters for the GC peak

If the experimental data for the sample were:

h = 4.2 mm
w = 0.2 mm

then:

$$\% p\text{-NCB} = (0.231)\ (4.2)\ (0.2)$$
$$= 0.19404 = 0.2$$

The certainty to which % *p*-NCB is measured is limited by the limit in the certainty to which the GC peak width is measured.

In final reporting of laboratory results, not only the rules applying to significant figures but the precision and accuracy of the method determine the number of figures reported. It is very time consuming for a laboratory supervisor to check all calculations to be sure your results and data can be interpreted properly. It is easy to see why technicians are only fully useful when they may be depended upon to provide data that do not have to be rechecked in too much detail.

Exponents

There are many calculations where the concept of significant figures cannot be applied. For example, the product $45 \times 2000 = 90{,}000$ cannot be rounded off using our rules because the product has no numbers to the right of the decimal. Exponents express numbers using

a decimal point, so the principles of significant figures can be used.

An exponent indicates the number of times a quantity is multiplied by itself. The numbers in the column on the left may be expressed exponentially as follows:

$$10 = 10^1$$
$$100 = 10 \times 10 = 10^2$$
$$1000 = 10 \times 10 \times 10 = 10^3$$
$$9000 = 9 \times 10 \times 10 \times 10 = 9 \times 10^3$$

Numbers less than 10 may also be written exponentially by using negative exponents:

$$10^{-n} = \frac{1}{10^n}$$

The numbers on the left may be expressed exponentially as follows:

$$0.1 = \frac{1}{10} = 10^{-1}$$

$$0.000000623 = \frac{6.23}{10^7} = 6.23 \times 10^{-7}$$

$$0.000000000000007 = 7 \times 10^{-15}$$
$$0.7050 = 7.050 \times 10^{-1}$$

When expressing a number with exponents, the decimal point is placed after the first numeral on the left

$$6251 = 6.251 \times 10^3 = 62.51 \times 10^2$$

Although all three expressions above are numerically equal, the middle one is in the correct form. If all numbers are written with only one numeral to the left of the decimal point, the rules for significant figures can be applied to give consistent precision in calculations.

You might correctly predict from the tabulation below that the exponent that is needed to express the number one exponentially is zero.

$$100 = 10^2$$
$$10 = 10^1$$
$$1 = 10^0$$
$$0.1 = 10^{-1}$$
$$0.01 = 10^{-2}$$

From the above examples, it is evident that writing numbers exponentially provides a precise method for expressing the correct number of significant figures, reduces the opportunities for errors, and also saves space.

Multiplication and Division with Exponents

In these exponential operations the number 10 is called the **base,** and each number of an operation is called a **term**. When multiplying terms with exponents, exponents having the same base are added. In division they are subtracted.

$$10^a \times 10^b \times 10^c = 10^{a+b+c}$$
$$\frac{10^a}{10^b} = 10^a \times 10^{-b} = 10^{a-b}$$

The following two examples illustrate the use of exponents. Such calculations, which are very common in chemistry, are much less tedious to work, and decimal places are easier to locate using exponential arithmetic.

$$1. \ \% \ Cl = \frac{N \ V \ E}{M \ D}$$

where

$$
\begin{aligned}
N &= 0.1235 \\
V &= 2.51 \ ml \\
E &= 35.453 \\
M &= 100.0 \ ml \\
D &= 1.237 \ g/ml.
\end{aligned}
$$

$$\% \ Cl =$$

$$\frac{(1.235 \times 10^{-1}) \ (2.51 \times 10^0) \ (3.543 \times 10^1)}{(10^2) \ (1.237)}$$

$$
\begin{aligned}
&= 0.0888 \\
&= 8.88 \times 10^{-2}
\end{aligned}
$$

2. Find $[H^+]$ in the following:

$$K_I = \frac{[H^+] \ [A^-]}{[HA]}$$

where:

$$
\begin{aligned}
K_I &= 1.7 \times 10^{-5} \\
[HA] &= 10^{-2} \ m/l \\
[A^-] &= 2 \times 10^{-1} \ m/l
\end{aligned}
$$

$$
\begin{aligned}
[H^+] &= \frac{K_I \ [HA]}{[A^-]} \\
&= \frac{(1.7 \times 10^{-5}) \ (10^{-2})}{(2 \times 10^{-1})} \\
&= 0.85 \times 10^{-6} \\
&= 8.5 \times 10^{-7}
\end{aligned}
$$

Logarithms

Just as exponential arithmetic is used as a calculation aid to obtain the location of the decimal, logarithms are used to perform tedious multiplications and divisions more simply by taking advantage of the additive property of logs in multiplication and their subtractive property in division.

The reason that logs may be used in this fashion is that a logarithm, or log, is an exponent. The common logarithm, or $\log_{10}$ is an exponent of the base 10. The $\log_{10}$ of 100 is 2 because $10^2 = 100$. In other words, **the log of a number is the power to which 10 must be raised to give that number.** An antilogarithm, or antilog, is the log changed back to corresponding number without exponents. Once numbers are expressed as logs, they are multiplied by adding the logs and divided by subtracting the logs, just as with any exponents with similar bases. Multiplications and divisions using logs consist of:

1. Expressing the numbers as logs
2. Performing the indicated additions and subtractions
3. Finding the antilog

Characteristics and Mantissas

Expressing a number as a log which is an even power of 10, such as 100, is a very simple matter. However, numbers which are not even powers of 10 require a log table. The first part of the log keeps track of the decimal and is called the **characteristic.** The second part of the number is called the **mantissa.** Consider the number 250. Using our exponential arithmetic, the number can be written 2.5×10^2. The characteristic for the log is 2 because the exponent of 10 is 2. The mantissa is found in the log tables by reading the first two numerals from the left-hand column and the third numeral from the top row. The mantissa for 250 is 0.3979. The log of the number 250 is written 2.3979. The characteristic of a log may be either positive or negative depending on the exponent of 10. The mantissa from the log table is always positive. The following examples will give you practice in working with logs.

1. To convert a number such as 4700 to its log equivalent, write the number exponentially:

$$4.7 \times 10^3$$

The characteristic is 3. Then look up 4.70 in the log table (Table I); find the mantissa, 6721. The log of 4700 is 3.6721.

2. To convert a number smaller than 1 such as 0.0000567 to a log, again write the number exponentially:

$$5.67 \times 10^{-5}$$

The characteristic is -5; look up the mantissa for 5.67 and find, .7536. The log is $\bar{5}.7536$. The bar over the characteristic shows that it is negative (but the mantissa is positive).

3. To find the antilog of a log, such as 3.6721, look up the mantissa, .6721, in the log table and find the number corresponding to it, which is 4.70. Then using the characteristic, write the number exponentially: 4.70×10^3. Finally, write the number as 4700. Similarly, the antilog of $\bar{2}.5051$ is 3.2×10^{-2} or 0.0320.

4. The following simple problem is worked first by ordinary arithmetic, then by exponential arithmetic, and finally by logs to demonstrate the properties of the different methods.

$$N_{NaOH} = \frac{N_{HCl} \, V_{HCl}}{V_{NaOH}}$$

where:

$N_{HCl} = 0.1000$
$V_{HCl} = 20.0$ ml
$V_{NaOH} = 10.0$ ml

(a) $N_{NaOH} = \dfrac{(0.1000)\,(20.0)}{(10.0)}$
$= 0.2000$

(b) $N_{NaOH} = \dfrac{(1 \times 10^{-1})\,(2.0 \times 10^1)}{(10^1)}$
$= 2 \times 10^{-1} = 0.2000$

(c) $N_{NaOH} =$ antilog (log 0.1000 + log 20.0) − log 10.0
$=$ antilog (1.000 + 1.3010) − 1
$=$ antilog (1.3010) = 0.200

The calculation in Example 4 shows the equivalence of the three approaches but fails to illustrate the usefulness of log calculations because the multiplication and division required was easy. Example 5 shows that a more complex calculation in mathematics is essentially no more complex when done with logs.

5. Calculate, using logs, % S from the following equation and data:

$$\% \, S = \frac{B \, F}{A \, W}$$

where:
B is the weight of barium sulfate; 0.973
F is the gravimetric factor; 137.35
A is the aliquot factor; 72.3
W is the sample weight; 8.7

Step 1. Find logs for the numbers

	Number	Log
B	0.973	$\bar{1}.9881$
F	137.35	2.1367
A	72.3	1.8591
W	8.7	0.9395

Step 2. Add the logs for multiplication; Subtract where division is required.

$$\begin{array}{c} \log B + \log F \\ \overline{1}.9881 \\ \underline{2.1367} \\ 2.1248 \end{array}$$

$$\begin{array}{c} \log A + \log W \\ 1.8591 \\ \underline{0.9395} \\ 2.7986 \end{array}$$

$$\begin{array}{c} \log BF - \log AW \\ 2.1248 \\ \underline{-2.7986} \\ \overline{1}.3262 \end{array}$$

Note: To multiply with logs when the sum of two mantissas reaches 1.0, the characteristic for the sum is increased by one. In division where the positive mantissa corresponding to the divisor is larger than the positive mantissa for the dividend, the characteristic for the dividend is algebraically reduced by one to keep track of the decimal.

Step 3. Find the antilog of $\overline{1}.3262$
    Answer: 0.212 % S

Note: Check your decimal by exponential arithmetic

$$\% \text{ S} \sim \frac{9.7 \times 10^{-1} \times 1.3 \times 10^{2}}{7.2 \times 10^{1} \times 8.7}$$

$$\sim \frac{10^{-1} \times 1.3 \times 10^{2}}{.72 \times 10^{2}}$$

$$\sim 2 \times 10^{-1} \sim 0.2$$

Use of Logs

Aside from the usefulness of logs in reducing the tediousness of a series of multiplications and divisions, they find use in many chemical relationships.

Calculation of [H^+]. The pH of solution is used to express the acidity of weakly acidic solutions. The log rela-

tionship of pH to the concentration of the hydrogen ion, [H^+] is illustrated in the following chart.

[H^+]	[H^+]	pH
.00000000001	10^{-11}	11
.0000001	10^{-7}	7
.001	10^{-3}	3
.01	10^{-2}	2

The relationship is written:

$$pH = \log \frac{1}{[H^+]} = -\log [H^+]$$

For an [H^+] concentration of 1.5×10^{-3} the pH lies between 2 and 3 and is calculated as follows:

$$\begin{aligned} pH &= \log \frac{1}{1.5 \times 10^{-3}} \\ &= -\log (1.5 \times 10^{-3}) \\ &= -(3.176) = -(-3+0.176) \\ &= -(-2.824) = 2.824 = 2.8 \end{aligned}$$

Calculation of [Cu^{2+}]. It is proposed that all the copper in a waste stream be removed by reduction with zinc metal.

$$Cu^{2+} + Zn^{\circ} \rightleftarrows Cu^{\circ} + Zn^{2+}$$

Using the equation,

$$E_1 - E_2 = \frac{0.59}{n} \log \frac{[Zn^{+2}]}{[Cu^{+2}]}$$

where:

$E_1 = +0.337$ V for $Cu^{2+} + 2\,e^- \rightarrow Cu^{\circ}$
$E_2 = -0.763$ V for $Zn^{2+} + 2\,e^- \rightarrow Zn^{\circ}$.

n = the number of electrons transferred in the reaction; *i.e.*, 2., calculate the Cu^{2+} concentration where excess zinc was added. The final Zn^{2+} concentration is $2 \times 10^{-2}M$.

$$\begin{aligned} E_1 - E_2 &= 0.33 - (-0.763) \\ &= \frac{0.059}{2} \log \frac{[Zn^{2+}]}{[Cu^{2+}]} \end{aligned}$$

$$1.09 = \frac{0.059}{2} \log \frac{2 \times 10^{-2}}{[Cu^{2+}]}$$

Table I. Log Table (1) (*Interpolation in this section is inaccurate)

N	0	1	2	3	4	5	6	7	8	9	1	2	3	4	5	6	7	8	9
10	0000	0043	0086	0128	0170	0212	0253	0294	0334	0374	*4	8	12	17	21	25	29	33	37
11	0414	0453	0492	0531	0569	0607	0645	0682	0719	0755	4	8	11	15	19	23	26	30	34
12	0792	0828	0864	0899	0934	0969	1004	1038	1072	1106	3	7	10	14	17	21	24	28	31
13	1139	1173	1206	1239	1271	1303	1335	1367	1399	1430	3	6	10	13	16	19	23	26	29
14	1461	1492	1523	1553	1584	1614	1644	1673	1703	1732	3	6	9	12	15	18	21	24	27
15	1761	1790	1818	1847	1875	1903	1931	1959	1987	2014	*3	6	8	11	14	17	20	22	25
16	2041	2068	2095	2122	2148	2175	2201	2227	2253	2279	3	5	8	11	13	16	18	21	24
17	2304	2330	2355	2380	2405	2430	2455	2480	2504	2529	2	5	7	10	12	15	17	20	22
18	2553	2577	2601	2625	2648	2672	2695	2718	2742	2765	2	5	7	9	12	14	16	19	21
19	2788	2810	2833	2856	2878	2900	2923	2945	2967	2989	2	4	7	9	11	13	16	18	20
20	3010	3032	3054	3075	3096	3118	3139	3160	3181	3201	2	4	6	8	11	13	15	17	19
21	3222	3243	3263	3284	3304	3324	3345	3365	3385	3404	2	4	6	8	10	12	14	16	18
22	3424	3444	3464	3483	3502	3522	3541	3560	3579	3598	2	4	6	8	10	12	14	15	17
23	3617	3636	3655	3674	3692	3711	3729	3747	3766	3784	2	4	6	7	9	11	13	15	17
24	3802	3820	3838	3856	3874	3892	3909	3927	3945	3962	2	4	5	7	9	11	12	14	16
25	3979	3997	4014	4031	4048	4065	4082	4099	4116	4133	2	3	5	7	9	10	12	14	15
26	4150	4166	4183	4200	4216	4232	4249	4265	4281	4298	2	3	5	7	8	10	11	13	15
27	4314	4330	4346	4362	4378	4393	4409	4425	4440	4456	2	3	5	6	8	9	11	13	14
28	4472	4487	4502	4518	4533	4548	4564	4579	4594	4609	2	3	5	6	8	9	11	12	14
29	4624	4639	4654	4669	4683	4698	4713	4728	4742	4757	1	3	4	6	7	9	10	12	13
30	4771	4786	4800	4814	4829	4843	4857	4871	4886	4900	1	3	4	6	7	9	10	11	13
31	4914	4928	4942	4955	4969	4983	4997	5011	5024	5038	1	3	4	6	7	8	10	11	12
32	5051	5065	5079	5092	5105	5119	5132	5145	5159	5172	1	3	4	5	7	8	9	11	12
33	5185	5198	5211	5224	5237	5250	5263	5276	5289	5302	1	3	4	5	6	8	9	10	12
34	5315	5328	5340	5353	5366	5378	5391	5403	5416	5428	1	3	4	5	6	8	9	10	11
35	5441	5453	5465	5478	5490	5502	5514	5527	5539	5551	1	2	4	5	6	7	9	10	11
36	5563	5575	5587	5599	5611	5623	5635	5647	5658	5670	1	2	4	5	6	7	8	10	11
37	5682	5694	5705	5717	5729	5740	5752	5763	5775	5786	1	2	3	5	6	7	8	9	10
38	5798	5809	5821	5832	5843	5855	5866	5877	5888	5899	1	2	3	5	6	7	8	9	10
39	5911	5922	5933	5944	5955	5966	5977	5988	5999	6010	1	2	3	4	5	7	8	9	10
40	6021	6031	6042	6053	6064	6075	6085	6096	6107	6117	1	2	3	4	5	6	8	9	10
41	6128	6138	6149	6160	6170	6180	6191	6201	6212	6222	1	2	3	4	5	6	7	8	9
42	6232	6243	6253	6263	6274	6284	6294	6304	6314	6325	1	2	3	4	5	6	7	8	9
43	6335	6345	6355	6365	6375	6385	6395	6405	6415	6425	1	2	3	4	5	6	7	8	9
44	6435	6444	6454	6464	6474	6484	6493	6503	6513	6522	1	2	3	4	5	6	7	8	9
45	6532	6542	6551	6561	6571	6580	6590	6599	6609	6618	1	2	3	4	5	6	7	8	9
46	6628	6637	6646	6656	6665	6675	6684	6693	6702	6712	1	2	3	4	5	6	7	7	8
47	6721	6730	6739	6749	6758	6767	6776	6785	6794	6803	1	2	3	4	5	5	6	7	8
48	6812	6821	6830	6839	6848	6857	6866	6875	6884	6893	1	2	3	4	4	5	6	7	8
49	6902	6911	6920	6928	6937	6946	6955	6964	6972	6981	1	2	3	4	4	5	6	7	8
50	6990	6998	7007	7016	7024	7033	7042	7050	7059	7067	1	2	3	3	4	5	6	7	8
51	7076	7084	7093	7101	7110	7118	7126	7135	7143	7152	1	2	3	3	4	5	6	7	8
52	7160	7168	7177	7185	7193	7202	7210	7218	7226	7235	1	2	2	3	4	5	6	7	7
53	7243	7251	7259	7267	7275	7284	7292	7300	7308	7316	1	2	2	3	4	5	6	6	7
54	7324	7332	7340	7348	7356	7364	7372	7380	7388	7396	1	2	2	3	4	5	6	6	7

N	0	1	2	3	4		5	6	7	8	9		Proportional Parts								
													1	2	3	4	5	6	7	8	9
55	7404	7412	7419	7427	7435		7443	7451	7459	7466	7474		1	2	2	3	4	5	5	6	7
56	7482	7490	7497	7505	7513		7520	7528	7536	7543	7551		1	2	2	3	4	5	5	6	7
57	7559	7566	7574	7582	7589		7597	7604	7612	7619	7627		1	2	2	3	4	5	5	6	7
58	7634	7642	7649	7657	7664		7672	7679	7686	7694	7701		1	1	2	3	4	4	5	6	7
59	7709	7716	7723	7731	7738		7745	7752	7760	7767	7774		1	1	2	3	4	4	5	6	7
60	7782	7789	7796	7803	7810		7818	7825	7832	7839	7846		1	1	2	3	4	4	5	6	6
61	7853	7860	7868	7875	7882		7889	7896	7903	7910	7917		1	1	2	3	4	4	5	6	6
62	7924	7931	7938	7945	7952		7959	7966	7973	7980	7987		1	1	2	3	3	4	5	6	6
63	7993	8000	8007	8014	8021		8028	8035	8041	8048	8055		1	1	2	3	3	4	5	5	6
64	8062	8069	8075	8082	8089		8096	8102	8109	8116	8122		1	1	2	3	3	4	5	5	6
65	8129	8136	8142	8149	8156		8162	8169	8176	8182	8189		1	1	2	3	3	4	5	5	6
66	8195	8202	8209	8215	8222		8228	8235	8241	8248	8254		1	1	2	3	3	4	5	5	6
67	8261	8267	8274	8280	8287		8293	8299	8306	8312	8319		1	1	2	3	3	4	5	5	6
68	8325	8331	8338	8344	8351		8357	8363	8370	8376	8382		1	1	2	3	3	4	4	5	6
69	8388	8395	8401	8407	8414		8420	8426	8432	8439	8445		1	1	2	2	3	4	4	5	6
70	8451	8457	8463	8470	8476		8482	8488	8494	8500	8506		1	1	2	2	3	4	4	5	6
71	8513	8519	8525	8531	8537		8543	8549	8555	8561	8567		1	1	2	2	3	4	4	5	5
72	8573	8579	8585	8591	8597		8603	8609	8615	8621	8627		1	1	2	2	3	4	4	5	5
73	8633	8639	8645	8651	8657		8663	8669	8675	8681	8686		1	1	2	2	3	4	4	5	5
74	8692	8698	8704	8710	8716		8722	8727	8733	8739	8745		1	1	2	2	3	4	4	5	5
75	8751	8756	8762	8768	8774		8779	8785	8791	8797	8802		1	1	2	2	3	3	4	5	5
76	8808	8814	8820	8825	8831		8837	8842	8848	8854	8859		1	1	2	2	3	3	4	5	5
77	8865	8871	8876	8882	8887		8893	8899	8904	8910	8915		1	1	2	2	3	3	4	4	5
78	8921	8927	8932	8938	8943		8949	8954	8960	8965	8971		1	1	2	2	3	3	4	4	5
79	8976	8982	8987	8993	8998		9004	9009	9015	9020	9025		1	1	2	2	3	3	4	4	5
80	9031	9036	9042	9047	9053		9058	9063	9069	9074	9079		1	1	2	2	3	3	4	4	5
81	9085	9090	9096	9101	9106		9112	9117	9122	9128	9133		1	1	2	2	3	3	4	4	5
82	9138	9143	9149	9154	9159		9165	9170	9175	9180	9186		1	1	2	2	3	3	4	4	5
83	9191	9196	9201	9206	9212		9217	9222	9227	9232	9238		1	1	2	2	3	3	4	4	5
84	9243	9248	9253	9258	9263		9269	9274	9279	9284	9289		1	1	2	2	3	3	4	4	5
85	9294	9299	9304	9309	9315		9320	9325	9330	9335	9340		1	1	2	2	3	3	4	4	5
86	9345	9350	9355	9360	9365		9370	9375	9380	9385	9390		1	1	2	2	3	3	4	4	5
87	9395	9400	9405	9410	9415		9420	9425	9430	9435	9440		0	1	1	2	2	3	3	4	4
88	9445	9450	9455	9460	9465		9469	9474	9479	9484	9489		0	1	1	2	2	3	3	4	4
89	9494	9499	9504	9509	9513		9518	9523	9528	9533	9538		0	1	1	2	2	3	3	4	4
90	9542	9547	9552	9557	9562		9566	9571	9576	9581	9586		0	1	1	2	2	3	3	4	4
91	9590	9595	9600	9605	9609		9614	9619	9624	9628	9633		0	1	1	2	2	3	3	4	4
92	9638	9643	9647	9652	9657		9661	9666	9671	9675	9680		0	1	1	2	2	3	3	4	4
93	9685	9689	9694	9699	9703		9708	9713	9717	9722	9727		0	1	1	2	2	3	3	4	4
94	9731	9736	9741	9745	9750		9754	9759	9763	9768	9773		0	1	1	2	2	3	3	4	4
95	9777	9782	9786	9791	9795		9800	9805	9809	9814	9818		0	1	1	2	2	3	3	4	4
96	9823	9827	9832	9836	9841		9845	9850	9854	9859	9863		0	1	1	2	2	3	3	4	4
97	9868	9872	9877	9881	9886		9890	9894	9899	9903	9908		0	1	1	2	2	3	3	4	4
98	9912	9917	9921	9926	9930		9934	9939	9943	9948	9952		0	1	1	2	2	3	3	4	4
99	9956	9961	9965	9969	9974		9978	9983	9987	9991	9996		0	1	1	2	2	3	3	3	4

$$36.9 = \log \frac{2 \times 10^{-2}}{[Cu^{2+}]}$$

$$10^{36.9} = \frac{(2 \times 10^{-2})}{[Cu^{2+}]}$$

$$[Cu^{2+}] = \frac{2 \times 10^{-2}}{10^{36.9}} = 2 \times 10^{-38.9}$$

$$[Cu^{2+}] = 2 \times 1.2 \times 10^{-39}$$

$$[Cu^{2+}] = 2.4 \times 10^{-39}$$

$$[Cu^{2+}] = 2 \times 10^{-39} \, M$$

The calculation shows that this is a very complete removal for most purposes of copper ion from the solution.

Calculation of Numbers with Exponents. A value may be calculated for a number raised to a power using logs. The following relationship is used:

$$\log a^n = n \log a$$

To calculate a value for $7.25^{3.27}$:

$3.27 \log 7.25 = (3.27)(0.8603)$
$3.27 \log 7.25 = 2.8131$
antilog of $2.831 = 677$
$\qquad 7.25^{3.27} = 677$

Calculation of Roots. Logs are used to calculate a root of a number by the following relationship:

$$\log \sqrt[n]{a} = \frac{1}{n} \log a$$

Calculate a value for:

$$\sqrt[3]{61}$$

$$\frac{1}{3} \log 61 = \frac{1.7853}{3}$$

$$= 0.5955$$

antilog of $0.5955 = 3.94$

$$\sqrt[3]{61} = 3.94$$

Analyzing the Problem

Now that you understand the concepts of significant figures, exponents, and logarithms, we must tackle the problem of using data to calculate solutions to experimental questions. For any computation the first step is to be sure you understand what is wanted. *What is the question?* Make sure the equation or method you intend to use is the right one. Be clear about what data are to be used. Is an answer needed right away, an hour later, or tomorrow? The time available may determine how you do the problem. What form of answer is needed: a table, graph, or other? When all these matters are clear, we are ready to start the calculations.

You may need to make only one calculation for a single experiment, or you may have many sets of similar data from which to make the same type of calculation. In either case, you may find that the data you have cannot be directly substituted into the correct equation. The equation may require time in *seconds* and your data may be given in *minutes* or the equation may require *centimeters* while the data are in *meters*. Simply plugging data values into an equation will often give a wrong answer even though all the arithmetic is correct. How does this happen?

Use of Dimensions

Almost every number represents something more than just the number itself: 5 *fingers*, 61 *cents*, 2.53 *grams*, etc. Here fingers, cents, and grams are the **dimensions** (or units) of 5, 61, and 2.53, respectively. Sometimes more dimensions are needed to describe something properly; *e.g.*, 3.7M really means 3.7 moles of solute per liter of solution, or (3.7 moles solute)/(1 liter solution).

The line in the fraction means "per." The number 1 is omitted in such expressions. So, 3.7M is the same as (3.7 moles of solute)/(liter of solution). You should be able to recognize the real dimensions of every number. Be careful when abbreviating. You should use g for grams, l. for liter, etc., but if there are two substances involved, be specific. Write "g of solute" or "g of solvent" or "g of solution" so that there can be no confusion.

The proper use of dimensions simplifies calculations and reduces errors. First you need to lay the groundwork by practicing the use of dimensions. This is one tool that leads to reliability because it answers such questions as "Do I multiply or divide?"

Almost every problem begins with given information including dimensions and requests an answer in specified dimensions. The problem frequently calls for converting the given dimensions to those required and includes any numbers that lead to the answer. Dimensions are converted with **conversion factors,** and solving problems with them is often called **dimensional analysis.** You are already aware that multiplying anything by 1 does not change its value. At this point, note that *all conversion factors are physically equal to one* although the numerical portion is not equal to one. If you want to convert centimeters to inches, you must first know that 1 cm is the same length as 0.394 in. or that 1 in. is the same length as 2.54 cm. We can describe these equalities with the equations; 1 cm = 0.394 in., and 1 in. = 2.54 cm. To make the first equation a conversion factor, we divide each side by 0.394 in.:

$$\frac{1 \text{ cm}}{0.394 \text{ in}} = \frac{0.394 \text{ in}}{0.394 \text{ in}} = 1$$

This means that we can multiply any length measurement by 1 cm./0.394 in. and not change its true value since it is really being multiplied by 1. Thus, 1 cm/0.394 in. is a conversion factor. Likewise,

$$\frac{0.394 \text{ in.}}{1 \text{ cm}}, \frac{1 \text{ in.}}{2.54 \text{ cm}}, \text{ and } \frac{2.54 \text{ cm}}{1 \text{ in.}}$$

are all also physically equal to 1.

In Experiment 1–1 of *Modern Chemical Technology,* we needed to place an ink spot 2.0 cm from the bottom of a strip of filter paper. How could we determine where to place the spot if only a foot ruler was handy? We would need to convert the centimeters wanted to the inches we could measure. *What are we given?* 2.0 cm. *What is required?* Distance in inches.

For greatest reliability, our procedure demands that we start with the information we have and work toward the dimensions we want. Let us begin by writing an equation that shows the question. In this case the question is, "How many inches equals 2.0 cm?"

$$\underbrace{? \text{ in.}}_{\text{Question}} = \underbrace{2.0 \text{ cm}}_{\text{Given}}$$

If we do not know a convenient conversion factor that includes cm and in., we can easily find one in a table of conversion factors such as the one in this book. We find that one inch equals 2.54 centimeters. Thus, the next step gives:

$$\underbrace{? \text{ in.}}_{\text{Question}} = \underbrace{2.0 \text{ cm}}_{\text{Given}} \times \underbrace{\frac{1 \text{ in.}}{2.54 \text{ cm}}}_{\substack{\text{A conversion} \\ \text{factor that} \\ \text{physically is} \\ \text{equal to 1}}}$$

We now see that the dimensions of cm will cancel and leave the wanted dimension, in.:

$$? \text{ in.} = 2.0 \text{ cm} \times \frac{1 \text{ in.}}{2.54 \text{ cm}} = 0.797$$

Applications of Dimensional Analysis

Conversion of Units. Although practically all laboratory measurements use the metric system, most production work uses the English system. A technician who is in a project that is going from laboratory scale to a production scale will usually find it necessary to convert all of his results into English units. Let us look at how one conversion might go.

A research project has found it necessary to use 300 ml of methanol as a solvent for a particular amount of reactant. This quantity must now be increased 1000 times and converted to gallons for a formulation to be given to the production department. Again, let us write an equation that describes the question and states the given information.

$$\underbrace{\text{? gal}}_{\text{Question}} = \underbrace{300 \text{ ml} \times 1000}_{\text{Given}}$$

(The value, 1000, is a dimensionless multiplier, or scale-up factor, to convert laboratory quantities to production quantities. Our only concern is whether this number will increase or decrease the size of our answer.)

Although a conversion factor from ml to gal may not be found, we can convert ml to liters and liters to gallons. We know that there are 1000 ml/l. and a conversion table tells us that there are 3.78 l./gal. Now we can write an equation that contains this information:

$$\underbrace{\text{? gal}}_{\text{Question}} = \underbrace{300 \text{ ml}}_{\text{Given}} \times \underbrace{1000}_{\substack{\text{Scale-up} \\ \text{factor}}} \times \underbrace{\frac{1.}{1000 \text{ ml}} \times \frac{3.78 \text{ l.}}{\text{gal}}}_{\text{Conversion factors}}$$

We see that unwanted dimensions on the right side are not cancelling; we have $l.^2/\text{gal}$. A look at the units in the last conversion factor shows that we need to invert it to get l. to cancel to give us the desired units:

$$\text{? gal} = 300 \text{ ml} \times 1000 \times \frac{1.}{1000 \text{ ml}} \times \frac{\text{gal}}{3.78 \text{ l.}}$$

Now we can do the necessary arithmetic:

$$\text{? gal} = \frac{300 \times 1000}{1000 \times 3.78} \text{ gal} = 79 \text{ gal}$$

Cost Determination. Technicians frequently need to order equipment and determine costs. How much would 5 funnels cost if a carton containing 72 funnels costs $77.80 delivered? Let us write an equation that describes the question and states the given information. In this case we want a final answer in dollars. The "5 funnels" must be used as a multiplier since more funnels means more dollars. So we can write:

$$\underbrace{\text{? \$}}_{\text{Question}} = \underbrace{\frac{\$77.80}{\text{carton}}}_{\text{Given}} \times \underbrace{\frac{\text{carton}}{72 \text{ funnels}}}_{\text{Given}} \times \underbrace{5 \text{ funnels}}_{\text{Given}}$$

(Try other arrangements of this information to see if you can get an answer with the dimensions of dollars.)

The arithmetic can now be done since all the dimensions cancel except dollars.

$$? \ \$ \ = \ \frac{\$77.80 \times 5}{72} \ = \ \frac{(\$7.78)\,(10)\,(5)}{(7.2)\,(10)} \ = \ \$5.40$$

(Note that we changed two of the numbers to exponential notation using powers of 10 (10^1 here). In this example, this was not very useful, but it is frequently extremely useful when working with both small and large numbers as was shown in the previous section on exponents.)

Regardless of the source of conversion factors—your memory, a table, or the problem statement—they must have an actual value of 1. When all the necessary conversion factors are present in the correct relationship, all units will cancel except those wanted in the answer. This condition provides the answer to the questions, "Have I used all the necessary conversion factors?" and "Do I multiply or divide?"

$$1 \ = \ \frac{2.54 \text{ cm}}{\text{in.}}; \ 1 \ = \ \frac{3.78 \text{ l.}}{\text{gal}}; \ 1 \ = \ \frac{\$77.80}{\text{carton}}$$

Percentage Calculations. In some instances a problem becomes easier to solve by restating it to describe the dimensions better. Percent problems really ask, "How many per hundred?" To say, "butter is 81% (by weight) butterfat," means, "There are 81 parts by weight of butterfat in 100 parts by weight of butter." We could say, "There are 81 g of butterfat per 100 g buttter." Percent is a decimal fraction multiplied by 100 and does not really have dimensions of its own.

What percent by weight is the silver in silver chloride? We have been given no numbers to work with, but we can use our knowledge of chemistry to reword the question and ask, "How many grams of silver (Ag) are there in 100 grams of silver chloride (AgCl)?" We also know that the atomic weight of Ag is 107.9, that the molecular weight of AgCl is 143.4, and that 1 mole of AgCl contains 1 mole of Ag. For convenience, we can use these weights in grams to give us the weight of a mole. With this information we can write the equation:

$$\underbrace{? \ \% \ \text{Ag}}_{\text{Question}} \ = \ \underbrace{? \ \text{g Ag in 100 g AgCl}}_{\substack{\text{Question restated} \\ \text{as above}}}$$

$$= \ \underbrace{100 \text{ g AgCl}}_{\substack{\text{Given as} \\ \text{defined above}}} \ \times \ \underbrace{\frac{107.9 \text{ g Ag}}{\text{mole Ag}}}_{\text{Given}} \ \times \ \underbrace{\frac{1 \text{ mole Ag}}{1 \text{ mole AgCl}}}_{\text{Given}} \ \times \ \underbrace{\frac{\text{mole AgCl}}{143.4 \text{ g AgCl}}}_{\text{Given}}$$

All undesired units cancel and the remaining arithmetic is:

$$? \ \% \ = \ ? \ \text{g Ag in 100 g AgCl} \ = \ \frac{100 \times 107.9}{143.4} \ \text{g Ag}$$

$$= \ \frac{(10^2)\,(1.079)\,(10^2)}{(1.434)\,(10^2)} \ \text{g Ag}$$

$$= \ 75.2 \text{ g Ag in 100 g AgCl}$$

Our answer can be properly stated as 75.2% Ag in AgCl.

Use of Mole Ratios. In chemistry, we have a frequently occurring problem that can be handled by using a ratio of moles of chemical species as we would use a conversion factor.

Let us calculate the percent Ag in Ag_2SO_4. The information we have is the atomic weight of Ag (107.9), the molecular weight of Ag_2SO_4 (311.7), and the knowledge that there are two moles of Ag in one mole of Ag_2SO_4. (For this example, we will use percentage in a more conventional way than in the above example. We will calculate a fraction and multiply by 100.) Now we can write our equation:

$$\underbrace{? \; \% \; Ag}_{\text{Question}} = \underbrace{\frac{g \; Ag}{g \; Ag_2SO_4}}_{\text{Question restated}} \times 100$$

$$= \underbrace{\frac{107.9 \; g \; Ag}{mole \; Ag} \times \frac{2 \; mole \; Ag}{mole \; Ag_2SO_4} \times \frac{mole \; Ag_2SO_4}{311.7 \; g \; Ag_2SO_4}}_{\text{Given information}} \times \underbrace{100}_{\substack{\text{Percent} \\ \text{multiplier}}}$$

The desired dimensions are obtained and only the arithmetic remains.

$$? \; \% \; Ag = \frac{107.9 \; g \; Ag \times 2 \times 100}{311.7 \; g \; Ag_2SO_4}$$

$$= \frac{(1.079)(10^2)(2)(10^2) \; g \; Ag}{(3.117)(10^2) \; g \; Ag_2SO_4}$$

$$= 69.2\% \; Ag$$

Note that the last dimensions are g Ag and g Ag_2SO_4 which were used for the "Question restated" to describe how percentage is actually calculated for this problem. This procedure is necessary for all percentage problems.

This simplified dimensional analysis can be used to solve many of the stoichiometric or titrimetric problems that occur in chemistry. We can apply the same procedure used earlier to cancel dimensions. In this case we use the balanced equations that describe the chemical reactions in which we are interested. We will use the ratios of moles of chemical substance as given by the balanced chemical equation in the same way we used dimensions. Although we have a very useful calculational technique, we must still know the chemistry in order to use it. Examples are provided by the following problems.

Use of Balanced Equations. Sodium hydrosulfite, $Na_2S_2O_4$, MW 174.10, is used in the paper industry for bleaching ground wood and decolorizing waste paper, rags, and kraft pulps. The hydrosulfite in a 0.2381 g sample of this material was changed to sulfate with hydrogen peroxide. The sulfate was precipitated by adding $BaCl_2$ to yield 0.6002 g. Calculate the % $Na_2S_2O_4$ in the sample. Reactions:

$$Na_2S_2O_4 + 3H_2O_2 \longrightarrow Na_2SO_4 + H_2SO_4 + 2H_2O$$

$$Na_2SO_4 + H_2SO_4 + 2BaCl_2 \longrightarrow 2BaSO_4 + 2NaCl + 2HCl$$

After we are sure we have the correct chemical equations, we can immediately start writing the equation:

$$\underbrace{? \% \ Na_2S_2O_4}_{\text{Question}} = \underbrace{? \ \frac{g \ Na_2S_2O_4}{g \ sample} \ x \ 100}_{\text{Question restated}}$$

$$= \underbrace{\frac{0.6002 \ g \ BaSO_4}{0.2381 \ g \ sample}}_{\substack{\text{Experimental} \\ \text{information}}} \times \underbrace{\frac{mole \ BaSO_4}{233.3 \ g \ BaSO_4}}_{(1)} \times \underbrace{\frac{mole \ Na_2SO_4}{2 \ mole \ BaSO_4}}_{(2)}$$

$$\times \underbrace{\frac{mole \ Na_2S_2O_4}{mole \ Na_2SO_4}}_{(3)} \times \underbrace{\frac{174.10 \ g \ Na_2S_2O_4}{mole \ Na_2S_2O_4}}_{(4)} \times \underset{\substack{\text{Percent} \\ \text{multiplier}}}{100}$$

Factors (1) and (4) are completely described molecular weights arranged to cancel unwanted dimensions or mole quantities. (2) expresses the ratio of moles of $BaSO_4$ produced to moles of Na_2SO_4 reacted. (3) expresses that ratio of moles of Na_2SO_4 produced to moles of $Na_2S_2O_4$ reacted. Thus, we have the arithmetic:

$$? \ \% \ Na_2S_2O_4 = ? \ \frac{g \ Na_2S_2O_4}{g \ sample} \times 100 = \frac{0.6002 \times 174.10 \ g \ Na_2S_2O_4 \times 100}{0.2381 \ g \ sample \ x \ 2 \ x \ 233.3}$$

$$= \frac{(6.002)(10^{-1})(1.741)(10^2)(10^2) \ g \ Na_2S_2O_4}{(2.381)(10^{-1})(2)(2.333)(10^2) \ g \ sample} = 94.0\% \ Na_2S_2O_4$$

Notice that dimensions are carefully indicated. Molecular weight can be conveniently used as grams/mole and molarity can be used as moles solute/liter solution. If dimensions are clearly labeled, volumetric and titrimetric problems are just as easily handled.

Volumetric Calculations. How many grams of oxalic acid (90.0 grams/mole) will be needed to prepare 2 liters of 0.100*M* solution? If we have the required knowledge for this problem, we can immediately write the equation:

$$\underbrace{? \ g \ H_2C_2O_4}_{\text{Question}} = \underbrace{2.1 \ soln \times \frac{0.100 \ mole \ H_2C_2O_4}{l. \ soln}}_{\text{Given}} \times \underbrace{\frac{90.0 \ g \ H_2C_2O_4}{mole \ H_2C_2O_4}}_{\substack{\text{Chemical} \\ \text{information}}}$$

$$= 2 \ x \ 0.100 \ x \ 90.0 \ g \ H_2C_2O_4$$

$$= (2)(1.00)(10^{-1})(9.0)(10) \ g \ H_2C_2O_4 = 18.0 \ g \ H_2C_2O_4$$

Molarity Calculations. Commercial acetic acid has a density of 1.05 g per ml and is 99.5% (by weight) acetic acid. Its formula is CH_3COOH, and its MW is 60.05 g per mole. What is the molar concentration of commercial acetic acid? We can start by writing the equation.

$$? \frac{\text{mole}}{\text{l.}} = \underbrace{\frac{99.5 \text{ g CH}_3\text{COOH}}{100 \text{ g soln}} \times \frac{1.05 \text{ g soln}}{\text{ml soln}}}_{\text{Given}}$$

$$\times \underbrace{\frac{1000 \text{ ml soln}}{\text{l. soln}}}_{\substack{\text{Conversion} \\ \text{factor}}} \times \underbrace{\frac{\text{mole CH}_3\text{COOH}}{60.05 \text{ g CH}_3\text{COOH}}}_{\substack{\text{Chemical} \\ \text{information}}}$$

$$= \frac{99.5 \times 1.05 \times 1000 \text{ mole CH}_3\text{COOH}}{100 \times 60.05 \text{ l. soln}} = 17.4M$$

Titrimetric Calculations. Titrimetry sometimes requires that we use several conversion factors. However, this is where dimensional analysis has its greatest potential for preventing errors.

1. If 5.00 ml of a solution of 0.20N H$_2$SO$_4$ were used to titrate a 5.23 g sample containing NaOH, what is the percent NaOH in the sample? First, change 0.20N to 0.10M H$_2$SO$_4$ and write the chemical reaction:

$$2 \text{ NaOH} + \text{H}_2\text{SO}_4 \longrightarrow \text{Na}_2\text{SO}_4 + 2\text{H}_2\text{O}$$

Using the information available from the balanced chemical equation, we can now calculate the percentage of NaOH as follows:

$$\underbrace{? \% \text{ NaOH}}_{\text{Question}} = \underbrace{\frac{\text{g NaOH}}{\text{g sample}} \times 100}_{\text{Question restated}}$$

$$= \underbrace{\frac{5.00 \text{ ml soln}}{5.23 \text{ g sample}} \times \frac{0.10 \text{ mole H}_2\text{SO}_4}{\text{l. soln}}}_{\text{Given}} \times \underbrace{\frac{\text{l. soln}}{1000 \text{ ml soln}}}_{\substack{\text{Conversion} \\ \text{factor}}}$$

$$\times \underbrace{\frac{2 \text{ mole NaOH}}{\text{mole H}_2\text{SO}_4} \times \frac{40.0 \text{ g NaOH}}{\text{mole NaOH}}}_{\text{Chemical information}} \underbrace{\times 100}_{\substack{\text{Percent} \\ \text{multiplier}}}$$

$$= \frac{(5.00)(0.10)(2)(40.0)(10^2) \text{ g NaOH}}{(5.23)(10^3) \text{ g sample}} = 0.765\% \text{ NaOH}$$

2. The reaction for titrating sodium oxalate with potassium permanganate in acid solution is:

$$5\text{Na}_2\text{C}_2\text{O}_4 + 2\text{KMnO}_4 + 8\text{H}_2\text{SO}_4$$
$$\longrightarrow 10\text{CO}_2 + 5\text{Na}_2\text{SO}_4 + \text{K}_2\text{SO}_4 + 2\text{MnSO}_4 + 8\text{H}_2\text{O}$$

When 0.2010 g of sodium oxalate (Na$_2$C$_2$O$_4$) is dissolved in dilute sulfuric acid, 40.75 ml of a potassium permanganate solution are required to reach an end point. What is the molarity of the potassium permanganate solution? Let us write the equation.

$$? \ \frac{\text{mole KMnO}_4}{\text{l. soln}} = \underbrace{\frac{0.2010 \text{ g Na}_2\text{C}_2\text{O}_4}{40.75 \text{ ml soln}}}_{\text{Given}} \times \underbrace{\frac{1000 \text{ ml soln}}{\text{l. soln}}}_{\substack{\text{Conversion} \\ \text{factor}}}$$

 $? \frac{\text{mole KMnO}_4}{\text{l. soln}}$ — Question

$$\times \underbrace{\frac{\text{mole Na}_2\text{C}_2\text{O}_4}{134 \text{ g Na}_2\text{C}_2\text{O}_4} \times \frac{2 \text{ mole KMnO}_4}{5 \text{ mole Na}_2\text{C}_2\text{O}_4}}_{\text{Chemical information}}$$

$$= \frac{(0.2010)(2)(10^3) \text{ mole KMnO}_4}{(40.75)(134)(5) \text{ l. soln}}$$

$$= 1.472 \times 10^{-2} \ \frac{\text{mole KMnO}_4}{\text{l. soln}} \quad \text{or } 1.472 \times 10^{-2}M$$

3. If a solution of iron (II) requires 41.50 ml of $0.02421M$ KMnO$_4$ to reach the end point, how many grams of iron (II) were in the solution? The reaction may be written as:

$$5\text{Fe}^{2+} + \text{MnO}^{4-} + 8\text{H}^+ \longrightarrow 5\text{Fe}^{3+} + \text{Mn}^{2+} + 4\text{H}_2\text{O}$$

We can write the equation:

$$? \text{ g Fe (II)} = \underbrace{41.50 \text{ ml soln} \times \frac{0.2421 \text{ mole KMnO}_4}{\text{l. soln}}}_{\text{Given}} \times \underbrace{\frac{\text{l. soln}}{1000 \text{ ml soln}}}_{\substack{\text{Conversion} \\ \text{factor}}}$$

Question — $? \text{ g Fe (II)}$

$$\times \underbrace{\frac{5 \text{ mole Fe (II)}}{\text{mole KMnO}_4} \times \frac{55.85 \text{ g Fe (II)}}{\text{mole Fe (II)}}}_{\text{Chemical information}}$$

$$= \frac{(41.5)(0.02421)(5)(55.85) \text{ g Fe}}{(10^3)} = 2.81 \times 10^{-1} \text{ g Fe (II)}$$

4. If it is found that 27.55 ml of $0.02020M$ KMnO$_4$ are required to titrate the H$_2$O$_2$ in a 2.055 g sample, what is the percent by weight of H$_2$O$_2$ in the sample? The reaction is:

$$5\text{H}_2\text{O}_2 + 2\text{KMnO}_4 + 3\text{H}_2\text{SO}_4 \longrightarrow 2\text{MnSO}_4 + \text{K}_2\text{SO}_4 + 5\text{O}_2 + 8\text{H}_2\text{O}$$

The equation can be written as:

$$? \ \% \ \text{H}_2\text{O}_2 = \frac{\text{g H}_2\text{O}_2}{\text{g sample}} \times 100$$

$$= \frac{27.55 \text{ ml soln}}{2.055 \text{ g sample}} \times \frac{0.02020 \text{ mole KMnO}_4}{\text{l. soln}} \times \frac{\text{l. soln}}{1000 \text{ ml soln}}$$

$$\times \frac{5 \text{ mole H}_2\text{O}_2}{2 \text{ mole KMnO}_4} \times \frac{34.02 \text{ g H}_2\text{O}_2}{\text{mole H}_2\text{O}_2} \times 100$$

$$= \frac{(27.55)(0.02020)(5)(34.02)(10^2) \text{ g H}_2\text{O}_2}{(2.055)(10^3)(2) \text{ g sample}} = 2.303\% \ \text{H}_2\text{O}_2$$

5. Free phosphoric acid can be determined in fertilizer by dissolving the sample in H_2O, filtering the residue and titrating the phosphoric acid with standard base. A 0.8865 g sample of fertilizer treated as indicated, required 16.62 ml of 0.1503M NaOH for titration. Calculate the % P_2O_5 in the sample.

$$? \% \ P_2O_5 \ = \ \frac{g \ P_2O_5}{g \ sample} \times 100$$

$$= \frac{16.62 \ ml \ soln}{0.8865 \ g \ sample} \times \frac{0.1503 \ mole \ NaOH}{1 \ l. \ soln} \times \frac{1 \ l. \ soln}{1000 \ ml \ soln}$$

$$\times \frac{1 \ mole \ H_3PO_4}{2 \ mole \ NaOH} \times \frac{1 \ mole \ P_2O_5}{2 \ mole \ H_3PO_4} \times \frac{142 \ g \ P_2O_5}{mole \ P_2O_5} \times 100$$

$$= \frac{(16.62)(0.1503)(142)(10^2) \ g \ P_2O_5}{(0.8865)(2)(2)(10^3) \ g \ sample} = 10.0\% \ P_2O_5$$

6. 50.00 ml of 0.20000M HCl were added from a pipet to a 0.4000 g sample of limestone and the flask gently swirled until the sample was dissolved. The excess HCl was then titrated with 0.1000M KOH, 36.00 ml being required. This procedure is commonly called a *back titration*. Calculate the % $CaCO_3$ in the limestone.

$$? \% \ CaCO_3 \ = \ \frac{g \ CaCO_3}{g \ sample} \times 100$$

$$= \frac{(50 \ ml \ HCl \times \dfrac{0.2000 \ mole \ HCl}{1000 \ ml \ HCl} - 36 \ ml \ KOH \times}{0.4000 \ g \ sample}$$

$$\frac{\dfrac{0.1000 \ mole \ KOH}{1000 \ ml \ KOH} \times \dfrac{1 \ mole \ HCl}{1 \ mole \ KOH})}{0.4000 \ g \ sample}$$

$$\times \frac{1 \ mole \ CaCO_3}{2 \ mole \ HCl} \times \frac{100.09 \ g \ CaCO_3}{1 \ mole \ CaCO_3} \times 100$$

$$= \frac{\left(\dfrac{50 \times 0.2000}{1000} - \dfrac{36 \times 0.1}{1000}\right)}{0.4000} \times \frac{100.09 \times 100 \ g \ CaCO_3}{2 \ g \ sample}$$

$$= \frac{(6.4)(100.09)(10^2) \ g \ CaCO_3}{(10^3)(0.4)(2) \ g \ sample} = 80.0\% \ CaCO_3$$

Notice that in the numerator of the first long fraction, we have converted both terms to "moles of HCl" before subtracting. Terms to be added or subtracted from each other must have the same units.

Calculation Tables

Let us assume that you know what is wanted from a given calculation, you have the right equation and good data and all conversions are correctly made. If only a single value is required, you do the necessary calculations to get a final answer, check it for accuracy, and report it.

Often the computations become very monotonous because you repeat the same calculation over and over. For example, you may have a series of trial runs to optimize the output of a process. You may have 6–25 sets of data, each to be treated in the same manner. For these situations, it is very helpful to set up a calculation table.

We will prepare a calculation table for a rather complex equation. Let us assume we have 14 sets of $P_{2\infty}$ and P_e values from a dielectric properties experiment. The job is to calculate the dipole moments using the formula:

$$\mu = \sqrt{\frac{9kT}{4\pi N} (P_{2\infty} - 1.1\,Pe)}$$

The term $9kT/4\pi N$ contains values that are constant for this experiment, so we can calculate it once and call it C. Then, we need to find $1.1 \times P_e$, subtract that product from $P_{2\infty}$, multiply the results by C and, finally, take the square root. The calculation table helps to keep these steps straight. A suitable table would have column headings as shown below.

Notes:

1. The first column (1) is just for identification—sample number, experiment number, time, or other variable. Make sure the identification can be understood by someone else. A table of numbers is useless if identification is not clear and complete.

2. Often it is easier to enter all values and do the necessary calculations for one column before moving on to the next one. This means that one operation is done many times in succession; you get a feel for it and will be less likely to make mistakes.

Now we fill in the known data. At this point the table might look like Figure 1. The circled letters point out important features that are often overlooked.

A = The equation used and its source
B = Units on numbers—helps find an error if you get the wrong ones
C = Source of data
D = Calculation of constant
E = Cross-reference to source of experimental data
F = Blank line for test case
(It is often useful to try a set of numbers for which you know the answer. In this example, it would check the value of C and the further steps in the table itself.)
G = Multiplication of all numbers in Column 2 by 1.1 means fewer calculational operations; this procedure gives similar savings for other columns.

Now proceed with the calculation, filling the table and getting the desired set of 14 dipole moments.

You are still not finished.

(1) Sample No.	(2) P_e	(3) 1.1 x (2) or 1.1 x P_e	(4) $P_{2\infty}$	(5) (4) − (3) $P_{2\infty} - 1.1\,P_e$	(6) C x (5) or $C(P_{2\infty} - 1.1\,P_e)$	(7) $\mu = \sqrt{(6)}$

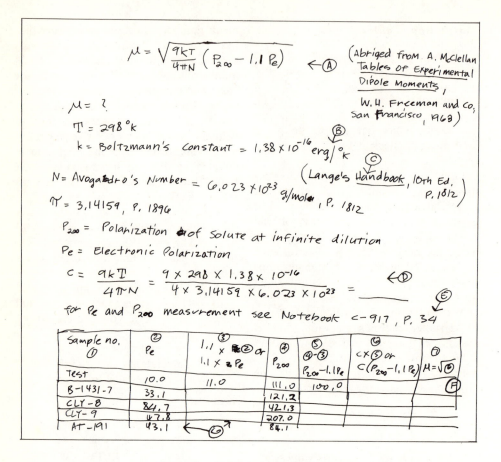

$$\mu = \sqrt{\frac{9kT}{4\pi N}\left(P_{2\infty} - 1.1\,P_e\right)} \quad \leftarrow Ⓐ$$

(Abriged from A. McClellan *Tables of Experimental Dipole Moments*, W. H. Freeman and Co, San Francisco, 1963)

$\mu = \;?$

$T = 298\,^{\circ}k$

$k = $ Boltzmann's Constant $= 1.38 \times 10^{-16}\,erg/^{\circ}k$ Ⓑ Ⓒ

$N = $ Avogadro's Number $= 6.023 \times 10^{23}\,g/mole$, P. 1812 (Lange's *Handbook*, 10th Ed., P. 1812)

$\pi = 3.14159$, P. 1896

$P_{2\infty} = $ Polarization of Solute at infinite dilution

$P_e = $ Electronic Polarization

$$c = \frac{9kT}{4\pi N} = \frac{9 \times 298 \times 1.38 \times 10^{-16}}{4 \times 3.14159 \times 6.023 \times 10^{23}} = \underline{\quad\quad} \quad \leftarrow Ⓓ \quad Ⓔ$$

for P_e and $P_{2\infty}$ measurement see Notebook $\leftarrow 917$, P. 34

Sample no. ①	② P_e	③ $1.1 \times$ ② or $1.1 \times 2 P_e$	④ $P_{2\infty}$	⑤ ④-③ $P_{2\infty}-1.1 P_e$	⑥ $c\times$⑤ or $c(P_{2\infty}-1.1P_e)$	⑦ $\mu=\sqrt{⑥}$
Test	10.0	11.0	111.0	100.0		Ⓕ
B-1431-7	33.1		121.2			
CLY-8	84.7		421.3			
CLY-9	47.8		207.0			
AT-191	43.1 $\leftarrow$ Ⓖ		84.1			

Figure 1.   Table before calculations

After Calculation

Two main tasks remain: checking the results and storing them so they can be found a week, a month, or a year from now. There should be no need to do a calculation over because results are lost. Enter the computation location in the laboratory notebook index, put down any references to supporting data, write the results on appropriate sample tags, mark and file the data sheets, instrument-drawn charts (spectra, chromatograms, etc.), punched tapes, or computer printouts. Keep in mind that the person wanting the results may not be you, so make sure that a stranger can find it. Don't waste the effort that produced the numbers by losing them in an inferior filing system.

Checking for Errors

Checking for errors is a little harder to describe because there are many kinds of errors. The first step is to become convinced that every calculation should be checked. All of us have a built-in belief that it is someone else who makes errors. Overcome that attitude and you will become known for reliable calculations.

As a first check ask yourself if the answer is *reasonable*. You should know the general range of expected answers. If you were calculating the price of a new car with certain accessories and you got $35,000 or $350.00, you would know something was wrong. In the same way, a solution that normally is about 50% benzene should not give an answer of 4.8% or 480%.

Another check is to compare the various values with each other. Again, your knowledge of the experiment can help you spot an error. Maybe all the results should be about the same. Are they? Or maybe there should be a trend such that as you go down the table the results should increase. Do they? This type of checking is another reason for being orderly in doing the calculations.

The next step is to get an approximate answer as shown in the following example.

Suppose a calculation looks like this (after dimensions have been properly handled):

$$\text{g needed} = \frac{43.71 \times 0.028 \times 4}{112.32}$$

To approximate the answer, round off each number and convert to exponential notation to get:

$$\text{g} \approx \frac{4 \times 10 \times 3 \times 10^{-2} \times 4}{10^2}$$

This can now be easily worked out:

$$\text{g} \approx 48 \times 10^{-3} \approx 0.05$$

The answer is about 0.05 g, not 0.5, 5, or 0.005. Note that whether you round up or down is unimportant as long as you are reasonable about it.

Approximation helps to check in two ways. It helps find the decimal point, which you must do anyway if you are using a slide rule, and it will reveal any major errors in the arithmetic. Decimal errors and wrong entries are among the most common mistakes, so approximation is a valuable check method.

Since decimal errors are so common, let us look at some guidelines. Adding and subtracting usually do not give much trouble. Align the decimals above each other and work column by column starting at the right and proceeding to the left. For multiplication (of two factors) there are three possibilities each leading to a different numerical "region" for the answer:

1. If both factors are greater than 1, the answer is larger than either factor.
2. If one factor is less than 1, the answer is between the two factors.
3. If both factors are less than 1, the answer is less than either factor.

For division (of two factors) there are two cases:

1. If the larger factor is on top, the answer is greater than 1.
2. If the smaller factor is on top, the answer is less than 1.

Another checking method is to redo the problem, but in a different order.

Addition: add the column from bottom up.

Subtraction: add the result to bottom figure to get top figure.

Multiplication: reverse order of the two factors and remultiply.

Division: multiply divisor by the result you got and add any remainder to get the number you were dividing into.

Methods of Calculating

In this section we will discuss several devices that may help you get the numerical answer. Our main point is to show the advantages of the various methods and to help you choose the best one for a specific job. Table II shows the methods and some aspects of them.

Table II. Calculation Methods

Method	Advantages	Drawbacks	Cost	Common Uses
Hand	Always available	Limited by your skill; limited to $+$, $-$, $\times$, $\div$; slow	0	Simple, rough, single calculations
Slide rule	Quick, readily available; can take logarithms; with some rules, can get powers and roots directly	Does not work for $+$, $-$; does not supply decimal point; is an analog device—accuracy limited to about 0.2%	$5–25	Single or multiple calculations; low accuracy
Logarithm tables	Turn $\times$ into $+$ and $\div$ into $-$; handles roots and powers easily; more accurate than slide rule	Requires new skill; does not work for $+$, $-$	$1.5–5	Single or multiple calculations; high accuracy
Table of values	Requires only correct "look-up"; fast	Only work for the particular computation used in deriving it; usually allows only one variable	0	Frequently used; fits simple and complex operations
Nomograph	Requires only correct "look-up"; fast; can accommodate several variables	Only for a given equation	0	Repetitious computation; low accuracy
Mechanical calculator	Requires only punching skill; fast and reliable	Limits depend on machine; simple ones are easy to operate but cannot do many things; more complex ones make it harder to set up problem but can handle more operations	$100–600	Frequently used; fits simple and complex operations
Programmable calculator	Need only punching skill; fast and reliable; storage features lesser repetitious punching of constant factors; programs can be used many times without redoing	More complicated than mechanical calculators needs programming skill	$2,000–10,000	Wide variety of jobs both simple and complex, tious
Electronic computers	High capacity, speed; high accuracy; great flexibility	Advanced programming skill, knowledge of some language	10^5 and up	Largest jobs, too costly unless used many times

Slide Rules. Slide rules are extensively used around laboratories. You may want to consider a round rule. There are two advantages. First, it is more compact than a straight rule and gives greater accuracy. An eight inch diameter rule has a 24-inch scale—more than twice that of a 10-inch straight rule. Second, a calculation never "falls off the end" of a circular rule. On a straight rule, 28×41 "falls off" if you put the left index on 28. It is then necessary to reset the right index on 28 to get 1148.

Logarithm Tables. Logarithm tables are available for two common bases: 10 and the number $e = 2.71828. \ldots$ Logarithms for base 10 are called **common** or **Briggsian** and the symbol *log* is used. Those using base e make up the **natural** or **Napierian** system and are denoted by *ln* or log_e. Tables commonly give 4, 5, or 6 decimal places. Natural logarithms require extensive tables because they cannot show the change of decimal place by simple integer changes in the characteristic as base 10 logarithms can. Natural logarithms are used in many equations that are derived by calculus. Usually they are handled by converting to base 10 logs. For any number N,

$$\ln N = 2.303 \log N$$

A table of values is often constructed by technicians for a specific, often-repeated job. During the construction particular care must be given to accuracy. An extra copy should be made so the table does not have to be redone when use spoils the first one.

Nomographs. A nomograph is a set of scales drawn to the proper sizes and with relative spacing so that a straight edge laid through two points gives the desired calculated value on a third scale. The construction of these graphs requires a known equation and adherence to certain rules. Though they are often used in the laboratory, nomographs are not usually prepared by technicians. They are much more difficult to prepare than tables, but they permit more complex calculations. The use of nomographs is explained in the next chapter.

Calculators. There are many types of machine calculators. Basically they provide for the four elementary arithmetical operations.

Programmable calculators are a new group of electronic machines. In addition to the four basic operations, they often include the ability to find log, ln, square root, and trigonometric functions —sin, cos, and tan. In addition, they have two other capacities not present in the mechanical machines. They can store some results or constants for later or repeated use. This means you can avoid a great deal of number entering. Programmable calculators have the ability to remember a program (a set of operations). These programs can be used over and over to save a tremendous amount of time and to avoid error. Most of these devices are easy to program. You do not need to learn a language or special set of commands. You need to get a physical picture of the way numbers and operational signals are stored, then analyze the problem into such steps. Some practice is required to do this well.

Computers. Electronic computers are powerful devices and are the ultimate in computing power. Chemical technicians seldom need to be computer specialists. However, they may need to get their data into and results out of computers. With the growth of time-sharing, there is an increasing chance you will have a terminal in your laboratory that is connected to a computer. Your job will be to get information from the experiments to the computer. This usually takes one of three forms: you type numbers into the terminal, you prepare punched cards, or you get punched paper tapes from the experiment and use them to transfer the information to the computer. The first two operations require typing skill. Typing ability may actually be more important than knowledge of programming languages.

Computers require communication in one of several rather restricted languages. FORTRAN (*formula translation*) and BASIC are the most common ones for laboratory applications. As the names imply, they were designed to make the computer language similar to mathematical equations. There are only 50 to 100 command words in each of these languages, but they provide tremendous flexibility. Some familiarity with these languages is helpful if you deal with electronic computers. There are several inexpensive books for beginners. Four good ones are:

1. Hellmut Golde, *Fortran IV and V for Engineers and Scientists,* The MacMillan Co., New York, 1966.
2. Daniel McCracken, *A Guide to FORTRAN IV Programming,*

John Wiley & Sons, Inc., New York, 1965.

3. Digital Equipment Co., *PDP–10 Timesharing Handbook* (Book 3), Maynard, Mass., 1970.

4. Hewlett-Packard Co., *2000B: A Guide to Time-Shared Basic,* Cupertino, Calif., 1970.

One important thing to remember about all computing devices is that they supply no judgment. They cannot tell if you have put in correct or incorrect numbers, whether the answer is of any use, or what is the next thing to do. Computer people have a saying: "garbage in, garbage out." The point is that the use of a computer, no matter how expensive or powerful, does not relieve you of the need for judgment. You must still ask, "Is this reasonable?"

Literature Cited

1. "Handbook of Chemistry and Physics," **54,** p. A-7, A-8, Chemical Rubber Co., Cleveland, 1973.

12

Use and Interpretation of Data

Much of the time consumed in laboratories is spent obtaining data. The interpretation and use of these data are the basis for many important decisions. As a technician, you may be reluctant to concern yourself with arithmetic, algebra, and statistics, but as you become more aware of the vital role your data and its interpretation play in your work, you will appreciate the need to understand some of the more basic concepts.

Calibration Plots

In the previous chapter under the section on significant figures, an equation was used to calculate % *p*-NCB.

$$\% \; p\text{-NCB} = k \; h \; w$$

You will use equations of this sort and their calibration plots routinely, and it is important to understand how to use these tools in presenting and interpreting data.

A technician in a gas chromatography (GC) laboratory for the Environmental School of Public Health of a university is assisting the city to determine whether the phenol concentration in the water effluent of a local refinery exceeds a variance permit maximum of 3 ppm. Standards are made up, and peak height increases with concentration. The sample from the refinery is 0.10 cm.

When these data are plotted to prepare a calibration plot, the result for the refinery sample may be read from the graph.

Phenol in Standard, (ppm)	Peak Height, (cm)
0	0
10	0.25
20	0.49
30	0.70

The first two graphs, Figures 1 and 2, show common errors in plotting data.

In Figure 1, the scale for the peak height is much too large. It is very

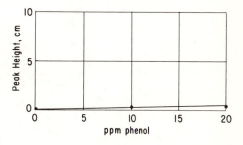

Figure 1.   GC response to phenol concentration

difficult to read corresponding peak heights accurately because the slope (rate of rise) of the line is too small to be interpreted easily. A second, less serious mistake is plotting data on graph

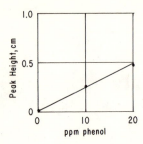

Figure 2. GC response to phenol concentration

paper which is difficult to read, as in Figure 2. Lack of subdivisions makes estimation of corresponding values very difficult. Figure 3 shows a correct plot of the data. It is easy to read that at 0.10 cm the phenol concentrations is 4.2 ppm and that the effluent exceeds the permit maximum of 3 ppm.

When plotted, not all data produce either a straight line or one that passes through the origin (point where both scales are 0).

Traces of mercury in waste water can be removed from water with a sulfide treatment. The mercury sulfide produced in this treatment is filtered through a sand bed, and the excess sulfide is air-oxidized to harmless, odorless com-

pounds. Assume that a continuous monitor is required to control feed of a sodium sulfide solution to assure contact of the contaminated effluent with 40 ppm excess sulfide ion. The laboratory supervisor has ordered a new ion-selective electrode which is supposed to respond very selectively to sulfide ion. You are asked to determine whether the electrode can measure sulfide ion for the range of 1–50 ppm sulfide. You prepare a series of standards, and using the new electrode, obtain millivolt readings as follows:

Sulfide ion, (ppm)	mv.
0	20
5	27
10	36
20	66
30	121
40	174
50	316

A plot of the data will produce a graph like that shown in Figure 4.

Corresponding values for both variables can easily be read from both straight and curve line graphs. However, interpretation of the data is simpler when a graph is a straight line. Often you can manipulate data to produce

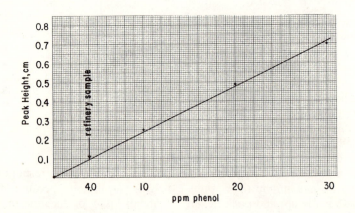

Figure 3. GC response to phenol concentration

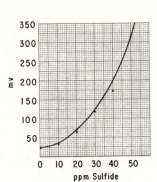

Figure 4. A calibration plot which is neither linear nor passes through the origin

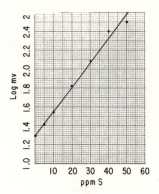

Figure 5. Plot of log mv vs. ppm S

such a graph. One way to do this is to convert the values for one of the variables to their logarithmic equivalents. When the logs of the mv values are plotted against ppm S, a straight line graph results (Figure 5).

Another way to reduce a curve like that in Figure 4 to a straight line plot is to use semi-log paper. Note that the mv data require two cycles on the log

scale to encompass the data, and the paper is therefore called 2-cycle semi-log paper (Figure 6). Many examples exist of laboratory data relationships which produce a straight line when plotted on semi-log paper (Figure 7). Other examples exist for curves which become straight lines when plotted on log-log paper.

Equations for Lines

The relationship of plotted data can be expressed as a simple equation if the graph is a straight line or a curve that can be transformed into a straight line by replotting a log relationship. Straight line calibration plots passing through the origin are expressed by the equation:

$$y = k x$$

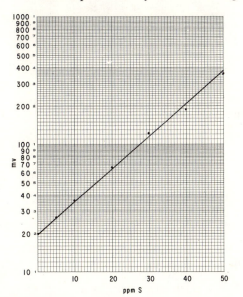

Figure 6. A 2-cycle semi-log plot of ppm S vs. mv

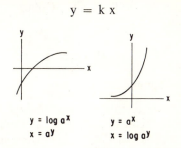

Figure 7. Appearance of data plots reducible to a straight line by a semi-log plot

where x and y are corresponding variables read from the graph, and k is the slope of the line.

The data for phenol on page 111 show that the GC peak height is proportional to phenol concentration. You can calculate an equation for this response in two ways. First, a single set of data may be used and substituted in the following equation:

$$\text{ppm phenol} = k\,h$$

where k is a constant (the slope of the line), and h is the peak height. At 30 ppm phenol, the peak height is 0.70 cm. k is calculated as follows:

$$30 \text{ ppm phenol} = k\,(0.70 \text{ cm})$$

$$k = \frac{30 \text{ ppm}}{0.70 \text{ cm}} = 42.8 \text{ ppm cm}^{-1}$$

$$= 43 \text{ ppm cm}^{-1}$$

The GC response factor, k, corresponds to the slope of the line.

To obtain the slope of a line passing through the origin, choose a point on the line, and divide its value on the vertical axes by the corresponding horizontal axis.

The second way to calculate the equation is to use values from the graph itself. Refer to Figure 3 and read a value for a point, *e.g.* 30 ppm phenol. The corresponding y-value of this point on the line is 0.72 cm. Obtain k by dividing:

$$k = \frac{30 \text{ ppm}}{0.72 \text{ cm}} = 41.6 \text{ ppm cm}^{-1}$$

$$= 42 \text{ ppm cm}^{-1}$$

The two calculated values for the GC response factor differ because in the first case, the actual data values were used while the second value was calculated from a point on the line. This line is the "best fit" for all the data, but it does not actually include some of the points. An equation which is calculated using the single point actually describes a line passing from that point to the origin while the line calculated from a data point may be slightly displaced from the line drawn using all the data. A slope average technique using all the data points may be used to calculate a better value.

Use all of the data in the table to calculate an average GC response.

ppm Phenol	Peak height (cm)	ppm cm⁻¹
10	0.25	40.0
20	0.49	40.8
30	0.70	42.9

column headers: ppm Phenol, Peak height (cm), ppm cm^{-1}

$$\text{Average} = \frac{40.0 + 40.8 + 42.9}{3}$$

$$= 41.2 = 41 \text{ ppm cm}^{-1}$$

Frequently, when data are plotted, the resulting graph does not pass through the origin. The equation for such a line is:

$$y = k\,x + b$$

where b is the value of y when x = zero. The equation for such a line may be calculated using the method shown above or that in the following example.

The following data were obtained for measurement of ppm iron in methanol by a colorimetric method.

ppm Fe	A
0	0.125
1	0.250
2	0.375
3	0.500

1. Write the equation for the line.

$$\text{ppm Fe} = k\,A + b$$

where k is the response factor, and b is the reagent blank.

2. Calculate the slope by reading two pairs of data points and calculating the rate of rise of the line.

$$k = \frac{Fe_2 - Fe_1}{A_2 - A_1} = \frac{(2-1)}{(0.375 - 0.250)}$$

$$= \frac{1}{(0.125)} = \frac{8 \text{ ppm Fe}}{A}$$

3. Substitute the value for the slope and using a set of data, solve for b.

$$2 \text{ ppm Fe} = (8 \text{ ppm Fe A}^{-1})(.375A) + b$$

$$b = 1.00 \text{ ppm Fe}$$

4. Check your result using another data point.

$$3 \text{ ppm Fe} = (8 \text{ ppm Fe A}^{-1})(0.500A) - 1.00 \text{ ppm Fe}$$

$$= 4 \text{ ppm Fe} - 1.00 \text{ ppm Fe}$$

The problem above showed how a reagent blank is calculated. When the standard used contained no added iron, a color developed with an absorbance, A, corresponding to that which would be caused by the equivalent of 1 ppm Fe.

Least Squares Fit

The convenience and precision gained by using an equation instead of a calibration plot is sufficiently great that most scientists try to calculate an equation to define the relationship for their data. Not all curves can be reduced readily to a straight line, but together with the semi-log and log-log reductions most algebraic equations used by chemists can be reduced to the simple linear equation for a straight line. In calculating the best line fit for data, a method called Least Squares Fit is used. In this method, the least variance (standard deviation squared) for the plotted variables is calculated for a line. In the case of a least squares data fit for a straight line, statisticians have developed a shortcut for making these somewhat tedious calculations.

The activity of a catalyst is measured in minutes by an activity test (time to reduce the hydrogen from 15 atm to 1 atm under standard test conditions) which is believed by the research director to be related to ppm sulfur on the catalyst. He has prepared a curve from the following data, and it is essentially a straight line.

S (ppm)	Activity (min)
49	104
48	105
100	106
101	107
150	107
148	108
198	108
199	109

The best equation for this line is in the format,

$$A = bS + c$$

where: A is the activity in minutes, S is ppm sulfur, and b and c are constants.

Table I. Calculations for Least Squares Fit Problem.

S(ppm)	S^2	$(S-\bar{S})$	A(min)	$(S-\bar{S})A$	$(S-\bar{S})^2$
49	2,401	−75	104	−7,800	5,625
48	2,304	−76	105	−7,980	5,776
100	10,000	−24	106	−2,544	576
101	10,201	−23	107	−2,461	529
150	22,500	26	107	2,782	676
148	21,904	24	108	2,592	576
198	39,204	74	108	7,992	5,476
199	39,601	75	109	8,175	5,625
Sum 993	148,115		854	756	24,859

It is calculated from minimizing variances by a least squares fit as follows:

1. Calculate ΣA, ΣS, ΣAS, ΣS^2 and ΣA^2. Use $S - \bar{S}$, where $\bar{S}$ is the mean for the sulfur analysis, and write the equation as follows:

$$A = b(S - \bar{S})$$

Arrange the data in tabular form and calculate the sums indicated in Table I.

The following sums are from the sum column of the table.

$$\Sigma S = 993$$
$$\Sigma S^2 = 148,115$$
$$\Sigma A = 854$$
$$\Sigma(S - \bar{S})A = 756$$
$$\Sigma(S - \bar{S}) = 24,859$$

The means, $\bar{S}$ and $\bar{A}$, are calculated using ΣS, ΣA, and N, the number of values.

$$\bar{S} = \frac{\Sigma S}{N} = \frac{993}{8} = 124$$

$$\bar{A} = \frac{\Sigma A}{N} = \frac{854}{8} = 107$$

2. Calculate the slope b

$$b = \Sigma(S - \bar{S})A / \Sigma(S - \bar{S})$$
$$= 756/24859$$
$$= 0.0304$$

3. Calculate the intercept, c

$$c = \bar{A} - b\bar{S}$$
$$= 107 - (0.0304)124$$
$$= 107 - 3.8$$
$$= 103.2$$

4. Substitute the values for c and b in the equation:

$$A = 0.0304\,S + 103.2$$

Fitting data by a least squares fit is somewhat tedious. For this reason it is often done with a computer. Normally, the data is first plotted, and depending on the shape of the curve obtained, least square fits for the constants in the equations are computed for all possible equation types.

A test for variable correlation is then made, again by using a computer, to determine which equation fits the data best in view of the constants which have already been calculated. Where the operator has no feeling for which equations could apply he may elect to try fits for all of the equation types in the computer's curve fitting program. Computers make these calculations so quickly that the cost is not prohibitive. However, the operator must evaluate the printout and decide which form shows the best correlation to the data.

If the calculations are not made by computer, they are generally made on a hand calculator. Alternatively, they may be made with logs to perform the multiplications and divisions required.

Nomographs

The nomograph or nomogram is a graph that allows you to read the value of a dependent variable when two or more independent variables are given. A straightedge is necessary for this calculation method.

Frequently a depressed boiling point of a compound is obtained under reduced pressure. The pressure, if measured, can be reported as a boiling point corrected to 760 torr, so other experimenters working at a different partial pressure can use this physical property for identification. On the nomograph in Figure 8, to find the boiling point corrected to 760 torr for a compound boiling at 100°C and 0.06 torr, line up a ruler with 100°C of the scale on the left with 0.06 torr of the scale on the right and read the value on the middle

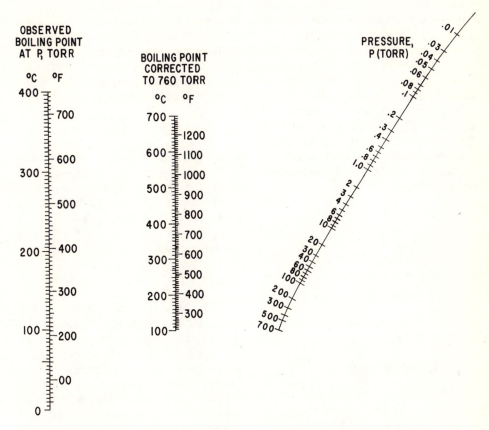

Figure 8. Nomograph calculation for solution to boiling point problems

scale where the line crosses the column. Your answer should be 342°C.

To calculate the °F boiling point for the compound at 1.0 torr, line up the ruler edge with 1.0 torr and 342°C from the boiling point corrected to 760 torr. Read your answer in the °F observed boiling point column. Your answer should be 297°F.

Solution of Equations with Two Dependent Variables

In both infrared and mass spectrometric analysis, it may not be possible to find a simple equation relating con-

centration to instrument response because two or more components may affect the response at the most sensitive wavelengths used. If two different wavelengths can be found where the two compounds show a different response when the responses are additive (which is normally the case), then two simultaneous equations can be written and solved.

Suppose you are to develop an infrared method for sterol F and B in mixtures. Pure sterol F shows the following relationships at 2930 cm^{-1} and 400 cm^{-1}.

$$\% \ F = k_{2930} \ A_{2930}$$
$$= k_{400} \ A_{400}$$

where $k_{2930} = 200$ and $k_{400} = 10$. Pure sterol B shows the following relationships at the same wavelengths.

$$\% B = g_{2930}\, A_{2930}$$
$$= g_{400}\, A_{400}$$

where $g_{2930} = 50$ and $g_{400} = 150$.

If an unknown sample measures an A_{2930} of 0.50 and an A_{400} of 0.70, the % B sterol in the mixture is calculated as follows. Assume no background corrections are needed and no interferences are present.

$$A_{2930} = \frac{\%F}{k_{2930}} + \frac{\%B}{g_{2930}}$$

$$\%B = 50\left(0.50 - \frac{\%F}{200}\right)$$

$$= 25 - \frac{\%F}{4}$$

$$A_{400} = \frac{\%F}{k_{400}} + \frac{\%B}{g_{400}}$$

$$\%B = 150\left(0.70 - \frac{\%F}{10}\right)$$

$$= 105 - 15\,\%F$$

$$= 1.7 - \frac{\%F}{4}$$

$$\%B - \frac{\%B}{60} = 23.3$$

$$0.834\,(\%B) = 23.3$$

$$\%B = 28\%$$

The Meaningless Nil

Frequently, in using equations of the type used to illustrate calculations from linear equations, a negative number is obtained. While the value for an iron analysis can be very small, it cannot be zero or negative. Some laboratories have resorted to reporting nil. A better approach is to calculate and report a "less than" value. Less than values may

be calculated from analyzing standards to obtain the method sensitivity from estimating limits of the equipment used, or from precision and accuracy statements for the determination, if available.

A GC method for ppm O_2 in liquid nitrogen uses the following equation:

$$\text{ppm } O_2 = \frac{(5.2)(mm)}{(cc)}$$

where 5.2 is a calibration constant, mm is the peak height, cc is the sample size and is 0.5 unless otherwise stated. The chart recording shows a quiver at the peak site, but the peak is less than 0.1 mm high. A less than ($<$) value for the sample can be estimated as follows:

$$\text{ppm } O_2 < \frac{(5.2)(0.1)}{0.5} < 1 \text{ ppm } O_2$$

From this calculation, the analyst can report the oxygen content is less than 1 ppm. If the oxygen content were 1 ppm, the peak height would have been a visible peak of 0.2 mm height. If the oxygen content were greater than 1 ppm, the peak height would have been greater than 0.1 mm. In general, less than values may be calculated by substituting the minimum readable value of the analog measurement into the equation used to calculate the amount present of the desired constituent. In the case of volumetric measurements the analog measurement is, for example, the readability of the burette. For gravimetric analysis, it is the readability of the balance. For an absorption spectrophotometer, it might be the smallest change on the meter which is readable.

Sensitivity

Calculation of the sensitivity of a method to allow reporting "less than values" and to make other meaningful comparisons has been based on instrument signal to noise ratio (S/N) where commonly the sensitivity is defined as 2S/N.

For example, Figure 9 shows an AAS recorder tracing where the noise level (baseline noise) is 0.02 units. Commonly, a signal less than twice the noise level is required from the instrument

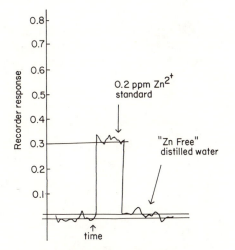

Figure 9. AAS recorder tracing for sample containing 0.2 ppm Zn^{2+}.

response to the measured element in the sample before detection of the element is certain. Signals smaller than 0.04 units, therefore, represent in this definition the readability of the instrument, and less than values and sensitivity are calculated as in the prior example by substituting 0.04 in the equation for that measurement. The following example illustrates this concept. (Another approach is based on the concentration of the analyte in the sample required to produce a 1% of full scale displacement in the recorder response.)

An AAS method for Zn^{2+} caused a displacement of 0.3 units out of a full scale response of 0–1 for a sample containing 0.2 ppm Zn^{2+}. The noise in this mode of operation was 0.02 units. Cal-

culate the method's sensitivity in terms of 2S/N as follows:

$$N = 0.02$$

The sensitivity of the method is measured by a signal twice N or 0.04 units. Note that the 0.2 ppm Zn^{2+} standard caused a meter displacement of 0.3 units. If in this AAS analysis meter, displacement is proportional to concentration, then one may write:

$$ppm\ Zn = k \times displacement$$

$$k = \frac{0.2}{0.3} = 0.66$$

A meter displacement then of 2N (0.04) corresponds to a sensitivity of 0.003 ppm Zn for the method.

$$ppm\ Zn = (0.66)(0.04) = .002_6 = .003$$

The form of the answer 0.002_6 may be unfamiliar to you. It is written in this manner to show that the product of (0.66)(0.04) should be rounded off to 0.003 and that the 6 of this product is an uncertain figure in terms of the measurement.

Precision—Repeatability and Reproducibility

The importance of knowing a method's precision and the general utility of expressing it in terms of standard deviation has resulted in widespread use of these expressions in transferring methods from one laboratory to another. With these expressions, the precision of different methods can be compared, (Figure 10), and the precision of a method can be defined in terms of repeatability and reproducibility.

A methods precision is measured by how well replicates for the same sample agree. The agreement between replicates which are obtained in the same

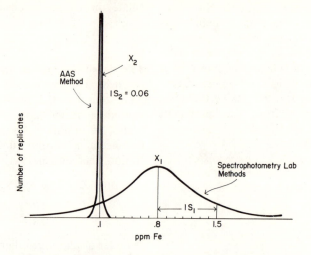

Figure 10. Comparison of two methods for measuring iron. The methods have different means ($\overline{X}_1$ and $\overline{X}_2$). The second method is more precise because it has a narrower range of probable values and a smaller single standard deviation, $1S_2$.

laboratory on the same equipment by the same operator measures the *repeatability* of a method. The agreement, or lack of it, between replicates from different laboratories on the same sample measures the *reproducibility* of the method.

Societies, such as the American Society for Testing and Materials (ASTM), standardize methods for many industries such as the rubber, cement, electrical, metal, ore, petroleum, building materials, etc. Industries' vendors and buyers base purchase contracts on a common method and always try to include statements of repeatability and reproducibility.

Confidence Levels

Any method showing normal scatter for the replicates on a particular sample will have the properties shown in Figure 15. The latter part of this chapter develops concepts such as mean, normal distribution, and standard deviation. These concepts are used in the interpretation of chemical data and deciding how "good" a number is in terms of confidence levels. Laboratories rarely have the luxury of obtaining more than a few replicates; usually only two or three. However, from even one result and a knowledge of the standard deviation, S, found in the published methods precision section, a probable statistical range can be calculated from the mean of the results for a sample. The equation below can be used to calculate the range for the 95% confidence level. The range may be calculated from a single determination or from the mean of a few replicates. The calculated value is interpreted to mean that 95 out of 100 replicates, if run for that sample, would fall in that range.

The equation for this calculation is:

Range of values (95% confidence level)

$$= \overline{X} \pm \frac{2S}{\sqrt{N}}$$

where X is the mean of the replicate for the sample, S is the standard deviation reported in the methods precision statement, and N is the number of replicates for the sample.

Calculate a range for the 95% confidence level if four replicates for % chloride in a sample were: 25.5, 24.5, 25.0, and 24.5%, and if the standard deviation for the method is 0.64% Cl.

Range (95% confidence level)

$$= \overline{X} \pm \frac{2S}{\sqrt{N}} = \overline{X} \pm S$$

$$\overline{X} = \frac{25.5 + 24.5 + 25.0 + 24.5}{4}$$

$$= 24.9\% = 24.9 \pm 0.64$$

$$= 24.3 \text{ to } 25.5\% \text{ Cl}$$

From this calculation, the analyst could say that if he ran 100 chloride analyses on this sample, 95 of them would be between 24.3 and 25.5% Cl. Most of the results as seen from scatter for the normal distribution curve would be close to the mean. Only a few results would be at the extremes of the reported range.

Accuracy

Chapter 15 discusses standard methods for analysis. Frequently, but not in the examples shown in Chapter 15, the publishers of methods will include a statement about the accuracy of the method, usually either in terms of standards or by comparison to referee methods.

The accuracy of a method is measured by how close a result obtained using that method approaches the true value. A method is normally standardized by analyzing standards (samples of known values) or by comparing results obtained to another "referee" method. Older established methods such as gravimetric or volumetric methods, which have been carefully studied by standardizing organizations and whose accuracy may be very well defined, are used to standardize newer, more rapid methods. Frequently, there is no method available which is accurate and suitable for evaluating a new method.

Often a technician will set up a method for the first time, which has been developed elsewhere. Standards are available from the National Bureau of Standards, standardizing organizations such as the ASTM, chemical supply houses, equipment-manufacturers, and companies offering special standards for this purpose.

Statistics: Distributions

Statistics is a useful tool for chemists and technicians. It establishes and checks quality control procedures, helps determine how good analytical data are, and describes submicroscopic phenomena (*e.g.*, the statistical model of atomic behavior). The application of most concern to you is establishing the reliability of your analytical data. Statistical techniques and terms have been used in previous sections. In this section we will explain and illustrate some statistical methods.

There are many good books which give detailed applications, definitions, and procedures of statistics. You are encouraged to read one of these books such as C. J. Brooks, *et al.*, "Mathematics and Statistics for Chemists," John Wiley & Sons. The treatment here will not be rigorous but rather an illustrative application using only elementary terms and definitions.

Table II. Results from Analysis of Geritol Tablets.

Sample	mg Fe Standard	mg Fe Geritol
1	298	51
2	298	49
3	295	47
4	296	47
5	293	46
6	299	49
7	302	53
8	300	50
9	303	52
10	304	54
11	311	59
12	289	44
13	299	51
14	298	49
15	299	48
16	297	47
17	300	51
18	303	52
19	304	54
20	302	51
21	301	50
22	299	49
23	296	47
24	295	46
25	315	60

A common laboratory experiment for chemistry students is to determine the amount of iron in Geritol. You know that no two tablets contain exactly the same amount of iron. However, you can obtain an estimate of the average amount of iron in Geritol tablets by running several determinations and calculating the arithmetic mean which is defined by the equation:

$$\text{Mean} = \frac{\Sigma x_i}{n}$$

where the symbol Σ means "the sum of," x_i is the ith determination (i = 1, 2, 3, ..., n), and n is the number of determinations. Stated in plain language, the mean of the determinations is equal to the sum of the individual determinations divided by the number of determinations.

The differences in the amount of iron in each tablet are caused by two factors, the true deviation from tablet to tablet and experimental error in determining the amount of iron. The latter factor, experimental error, is the reason iron standards are run along with the analysis of the tablets. The results of the standard analysis provide a check on your procedure and technique.

Suppose the results in Table II were obtained in the analysis of individual Geritol tablets and standard iron samples (300 mg Fe).

The first step in your analysis of the data is to construct a frequency distribu-

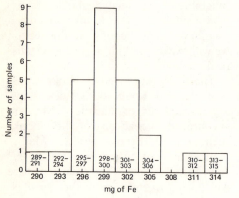

Figure 11. mg of Fe in standards

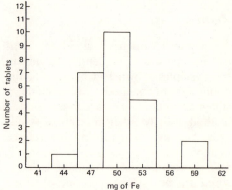

Figure 12. mg of Fe in Geritol

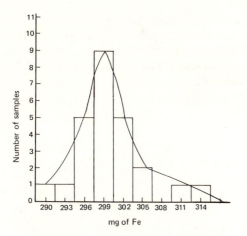

Figure 13. mg of Fe in standards

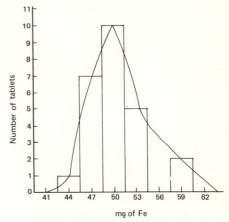

Figure 14. mg of Fe in Geritol

tion like the ones illustrated in Figures 11 and 12.

If you draw a smooth curve through the midpoints of the groups in Figures 11 and 12, you obtain a curve, somewhat distorted, which resembles a bell (Figures 13 and 14). These curves are approximations of what is called a normal curve resulting from a *normal distribution*. A normal distribution has the following properties (*see* Figure 15):

1. Approximately 68% of the distribution lies between the mean, $\overline{X}$, plus one standard deviation, S (or $\overline{X} + 1S$), and the mean minus one standard deviation (or $\overline{X} - 1S$).

2. Approximately 95% of the distribution lies between $\overline{X} + 2S$ and $\overline{X} - 2S$.

3. Approximately 99.7% of the distribution lies between $\overline{X} + 3S$ and $\overline{X} - 3S$.

The *standard deviation* is an estimate of the spread of a distribution. It is defined by the equation:

$$S = \sqrt{\frac{(x_i - \overline{X})^2}{n}}$$

In other words, you would substract the mean from each determination, square that difference, add all of the squared differences, divide the sum by the number of determinations, and take the positive square root of the result. This positive square root is the standard deviation.

If you assume the distributions of the data are normal, you can calculate the mean and standard deviation of each distribution and use the S as a guide in judging the accuracy of each of your determinations. When you perform the

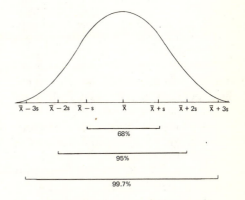

Figure 15. The normal curve

calculations on the data obtained from the standards, you obtain

$$\overline{X} = \frac{\Sigma(x_i)}{n} = \frac{7494}{25} = 299.76$$

$$S = \sqrt{\frac{\Sigma(x_i - \overline{X})^2}{n}} = \sqrt{\frac{688.56}{25}}$$

$$= \sqrt{27.54} = 5.25$$

Now you can look at the data for individual determinations in terms of the mean and standard deviation. You can arbitrarily establish the criteria for judging the analyses as: results which fall within the limits $\overline{X} \pm \frac{1}{2}S$ are very good, those results outside these limits but within the limits $\overline{X} \pm 1S$ are satisfactory and those results falling outside these limits are judged unreliable and unacceptable. Thus, you would obtain the boundaries illustrated in Figure 16.

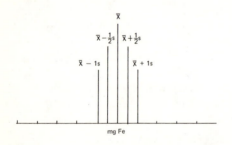

Figure 16. Setting boundaries for acceptable data for Fe in Geritol.

Then all results between 302.5 mg and 297.25 mg would be very good, those between 305 mg and 294.75 mg would be acceptable, and all the results outside these limits are unsatisfactory. You can then say that the analyses of standard iron samples 1, 2, 6, 7, 8, 13, 14, 15, 16, 17, 20, 21, 22 were very good; samples 3, 4, 9, 10, 18, 19, 23, 24 are acceptable, and samples 5, 11, 12, 25 are unacceptable.

You can now use these results on the standards to judge the analyses of the Geritol tablets. When you make the calculations on the Geritol results, you obtain:

$$\overline{X} = \frac{\Sigma x_i}{n} = \frac{1256}{25} = 50.24$$

$$S = \sqrt{\frac{\Sigma(x_i - \overline{X})^2}{n}} = \sqrt{\frac{344.56}{25}}$$

$$= \sqrt{13.78} = 3.71$$

Then $\overline{X} \pm \frac{1}{2}S = 52.1$, $\overline{X} - \frac{1}{2}S = 48.4$; $\overline{X} + 1S = 54.0$, $\overline{X} - 1S = 46.5$. Thus, analyses of samples No. 1, 2, 6, 8, 9, 13, 14, 17, 18, 20, 21, 22 are within $\overline{X} \pm \frac{1}{2}S$; samples No. 3, 4, 7, 10, 16, 19, 23 are within $\overline{X} \pm 1S$, and samples No. 5, 11, 12, 24, 25, are outside the limits. Now you must decide which analyses are acceptable. Consider Geritol sample No. 25 (60 mg), which is well outside the limits $\overline{X} \pm 1S$. You must decide if this deviation is caused by a true difference in iron content of the tablet or if the deviation is from experimental error. When you look at standard iron sample No. 25, which was run concurrently with the Geritol tablet sample No. 25, you see that the standard yielded an unacceptable result, implying considerable error. Therefore, you must assume that the Geritol sample No. 25 result also contains an unacceptable error. Now consider Geritol sample No. 24 which is also outside the $\overline{X} \pm 1S$ limits. When you look at standard iron sample No. 24, you see that its analysis was within acceptable limits. Hence, you should conclude that Geritol sample No. 24 is an acceptable estimate of iron in the tablet and the deviation is caused by a real difference in iron content of the tablet. You could continue our analysis by comparing standard and Geritol results for each sample.

You will note that the words "assume" and "imply" were used frequently, and the word "prove" was completely avoided. Statistical analyses prove nothing; they are only a tool to help observers infer relationships using their knowledge. In other words, they help an observer to make his best educated guess.

13

Drawings and Diagrams

Few individuals work alone in industrial, governmental, or university laboratories. Generally a team effort is required to conduct an efficient and effective laboratory operation. The team members may include a chemist, chemical technician, glass blower, machinist, and management personnel. An essential part of the laboratory operation is communication between members of the team.

Communications within a team, which may have its members physically separated, largely consist of verbally sharing ideas. However, the chemical technician often prefers to communicate his ideas and requests for assistance in writing or with sketches. Where instrumentation or other apparatus is involved, a technician frequently must read schematics of electrical apparatus, follow assembly drawings, or interpret instructions and sketches.

This chapter has been designed to acquaint a chemical technician with the terms and conventions used to express ideas in graphic form. After concluding this chapter, you should be able to read and interpret schematics, block diagrams, and mechanical drawings. Practical suggestions are given on how to prepare your own sketches from which apparatus can be built.

Types of Drawings

Mechanical. For chemical technicians there are essentially two ways that objects are represented with mechanical drawings. In one method, enough separate views are drawn to show clearly all of the details of the object. In a second method, a pictorial representation shows the object approximately as it would be seen by the eye. Basic types of drawings are illustrated in Figures 1 through 5.

Electrical. Every laboratory contains many electrical and electronic items ranging in complexity from simple variable voltage transformers to complex instrumentation such as recording spectrophotometers. The purpose of this unit is not to train the chemical technician to interpret complex drawings but to acquaint him with the basic symbols used to signify certain component functions. Good practices are outlined for sketching electrical circuits and using standard conventions to simplify representation of basic circuit functions. Schematic and block diagrams are shown in Figures 6 and 7.

Reading Drawings and Schematics

A drawing is read by visualizing details from the views prepared and men-

Figure 1. **One-view drawings.** *These are used when more than one view is unnecessary, for example, when a part is made of thin material such as plastic sheets or sheet metal.*

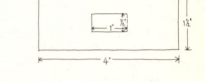

Figure 2. **Two-view drawings.** *Cylindrical apparatus generally requires only two views to show all details clearly.*

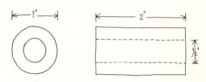

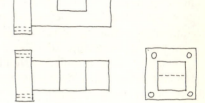

Figure 3. **Three-view drawings.** *Most scientific apparatus is sufficiently complicated to require at least three views to represent their form clearly. Auxiliary views are occasionally required to show hidden details.*

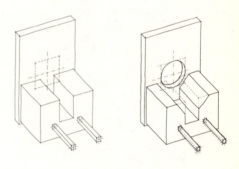

Figure 4. Pictorial drawings show an object approximately as it would be seen by the eye. **Isometric drawings** *or sketches are frequently prepared by chemical technicians to accompany requests for equipment to be fabricated by machinists, carpenters, or glassblowers.*

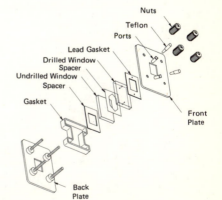

Figure 5. **Exploded drawings** *are used to illustrate the assembly method of a number of parts.*

tally orienting and combining the details finally to interpret the whole object. It should be clear that it would be an unnecessary and complex procedure to prepare drawings which would form a complete pictorial view. For example, consider the drawing in Figure 8. A visible cylindrical opening is seen at the top of the block. An understanding of mechanical drawing perspective indicates that there must be a continuation of the hole into or through the block. We mentally accept the fact that the dotted lines represent an extension of the hole into the block. With practice, conventions used to depict hidden details or size and shape are accepted when reading a well-prepared drawing.

The chemical technician should be familiar with the conventions used to represent dimensions so that he can interpret drawings correctly. In isometric drawings, diagonal lines are foreshortened and should not be accepted as an actual dimension. A three-view or orthographic mechanical drawing can be accurately drawn to scale. When reading or preparing drawings, keep in mind the degree of accuracy implied or stated for the drawing. For sketches, proportions are not meticulously maintained, but dimensions should be complete and accurate. Drawings made to scale should be used only when necessary, such as for building a prototype of a device where physical relationships need to be checked.

Dimensions may be marked on a drawing in two fashions: (1) as overall dimensions of the object, and (2) between points or surfaces associated through their functional relationships with adjoining parts (Figure 9). The best dimensions are those to which the craftsman can work to in producing the device. Here a knowledge of materials and capabilities of shop tools is essential. The next chapter provides an illustrated glossary of equipment fundamentals and material descriptions.

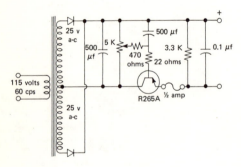

*Figure 6. A **schematic diagram** shows the electrical functions and connections for each component in a circuit. Numerical values of components are given for resistance, capacitance, voltages specified for transformers, current ratings for fuses, and reference designations for vacuum tubes and solid state devices, e.g., transistors and rectifiers.*

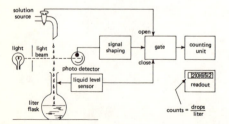

*Figure 7. **Block diagrams** convey basic information about the operation of a system of circuits or interrelating components. Much detailed information is omitted or encapsuled in block diagrams, thus simplifying the representation of complex apparatus. Multiconductor connections are represented by a single line in block diagrams.*

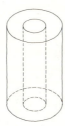

Figure 8. KBr die.

*Figure 9. Examples
of dimension lines.*

Pictorial Sketching Techniques

After developing a mental picture of the device to be drawn, the first step is to select the type of drawing. Since the details of one-, two-, and three-view drawings have been described briefly, we will discuss the method of perspective (isometric) drawing. Freehand isometric drawings will be sufficient for most of the needs of the chemical technician. Good drawings on plain paper require considerable experience, but effective freehand drawings can be easily constructed. Graph paper with lines ¼ in. apart on a 10 × 10 grid will serve nicely for most cases. The lines on the paper aid greatly in making projections, keeping the right proportions, and drawing straight and accurate lines. Sketching techniques are illustrated in Figures 10-17.

Several copies of a drawing will frequently be required for different people and the files. Make certain that the graph paper selected is printed in a color which will reproduce on the available office copier.

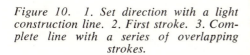

Figure 10. 1. Set direction with a light construction line. 2. First stroke. 3. Complete line with a series of overlapping strokes.

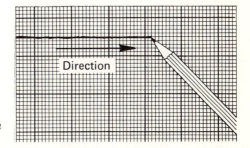

Figure 11. Sketch horizontal lines from left to right.

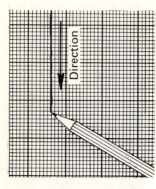

Figure 12. Sketch vertical lines from top to bottom.

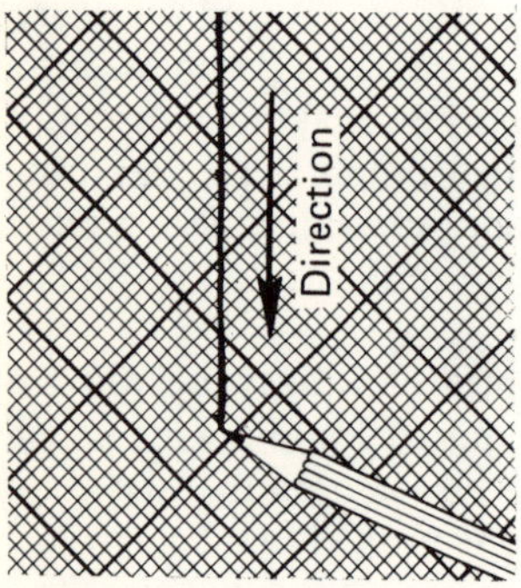

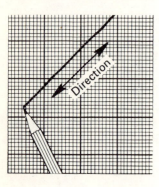

Figure 13. Turn the paper to sketch an inclined line as a vertical line. This is often a great help because of the awkward position of the inclined line.

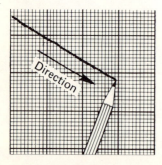

Figure 14. An inclined line sloping downward from right to left may be drawn in either direction.

Figure 15. An inclined line sloping downward from left to right is the most awkward position of any line. The paper should be turned as in Figure 16 to help draw a smooth, accurate line.

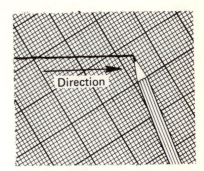

Figure 16. Paper turned to sketch an inclined line as a horizontal line. This should be done especially for the type of line shown in Figure 15.

Techniques and Tools

Several tools will be helpful for making drawings or sketches of equipment. These include: a 12 in. metal edge ruler, 6 or 8 in. 30° and 45° triangles, flexible curves, erasers and erasing shield, masking tape, a compass, and pencils with a relatively hard (No. 3) lead.

Mechanical Drawings. When starting a drawing, use masking tape to hold the paper on a smooth writing surface such as a clip board. Carefully choose the direction in which the object is to be viewed. Orient the object so that two principal faces are clearly shown. Be sure that important features will not be hidden. The next step is to draw the axes. In isometric drawings the three axes should be located from 120° from each other (Figure 17)—one vertical and two at 30° to the horizontal. Almost without exception the first lines drawn should be those that box in the object. Do not make the drawing too small. Observe carefully the following conventions: (1) all vertical lines must be parallel to the vertical axis throughout the drawing; (2) transverse lines must be parallel or converging; and (3) the axes must be kept flat to avoid distortion. Some useful drawing techniques are shown in Figures 18–24.

Figure 17. Pictorial axes for isometric sketching.

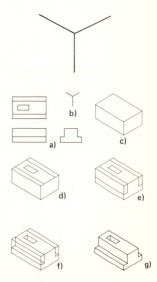

Figure 18. Steps in making an isometric sketch. a) Make orthographic drawing; b) draw axes; c) block in the enclosing shape; d) and e) draw outline of detail on top, front, and side; f) and g) finish by completing surfaces represented on the orthographic drawing.

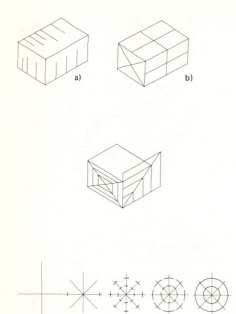

Figure 19. a) Frequently, faces must be divided into sections. This can usually be done satisfactorily by eye. b) Centers must often be located. Either use diagonals or judge the position of center lines.

Figure 20. The diagonals of a square or rectangle can be used to increase or decrease the rectangle symmetrically about the same center and in proportion. The grid system can also be used to increase or decrease size by equal units.

Figure 21. Drawing or sketching circular features presents certain problems. Freehand circles can be drawn by marking the radius on each side of the center lines. A more accurate method is to draw two diagonals in addition to the center lines and mark points equidistant from the center. The circles can then be drawn.

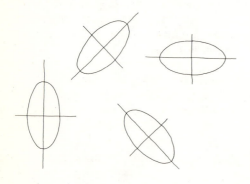

Figure 22. A circle in isometric drawings is an ellipse with a major diameter that is perpendicular to the rotation axis.

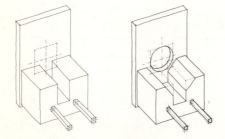

Figure 23. It is possible to box or draw an enclosing pictorial square with center lines to obtain the proper shape for circular forms or openings in equipment drawn in isometric form. When done properly, this assures correct shape and proportion.

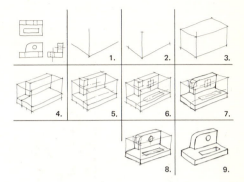

Figure 24. To review, note the steps needed to make a pictorial sketch. Progressive layout, (1) to (6); finishing, (7) to (9).

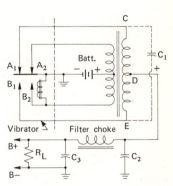

Figure 25. Schematic of a power supply.

Electrical Drawings. When reading a mechanical drawing, you can form a mental dimensional picture of a device. When reading an electrical schematic, you make no attempt to relate the drawing to the physical size or location of the components. The schematic is simply a diagrammatic representation of the function and connections of a circuit (Figure 25). In many cases schematics will be accompanied by one or more block diagrams which can be used to orient you to the overall functions of the apparatus, as well as to the functions of components within the apparatus. A typical example might be a block diagram of a complex recording spectrophotometer, accompanied by a block diagram of the optical system and the various circuits within the electrical detecting, amplifying, and readout sections by function. Block diagrams are an efficient way of introducing the reader to complex systems and should always be carefully read and understood before reading schematics or more complex mechanical diagrams. In troubleshooting, block diagrams can be effective for relating general functions of an instrument to a specific detail in the schematic or mechanical drawing. Use should be made of the enclosure, a drawing feature in which dotted lines are used in the complex diagram to identify a particular area with the function represented in a block diagram. For more details, *see* Figures 26 and 27 on the next three pages.

Figure 26. Diagrams for the components of a fluorescence spectrometer (see Note on page 136)

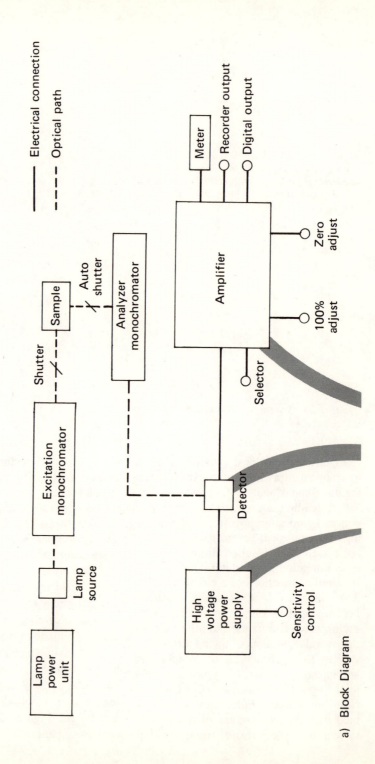

a) Block Diagram

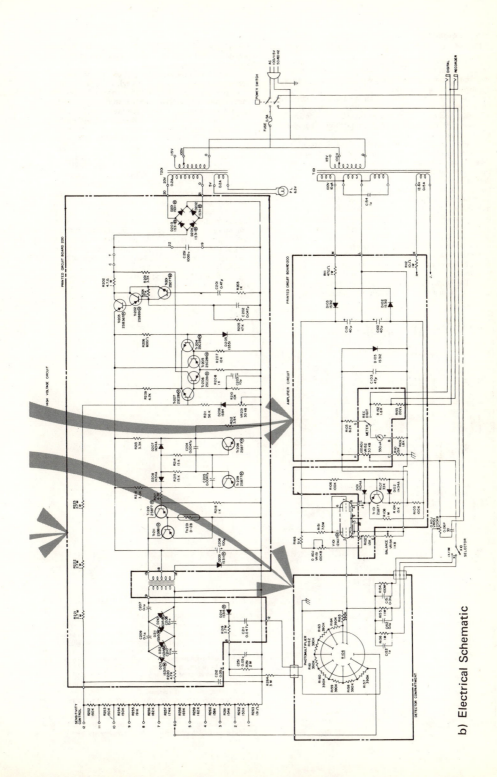

b) Electrical Schematic

Note for Figure 26
A Guide for Correlating Block Diagrams and Schematics

There is a tendency to regard instruments as "black boxes" that allow you to put a sample in and get an answer out. However, there are always problems of putting an instrument into operation, modifying existing equipment, increasing sensitivity, determining how to prevent inaccurate measurements or automating a process or method. Accurate information about laboratory instruments is usually supplied by the manufacturer in an operation manual.

Figures 26 a and b show a block diagram for a fluorescence spectrometer and an electrical schematic taken from an operational manual. Figure 27 gives the accepted symbols for basic electronic components and illustrates representative types. The purpose of these figures is to show the kinds of resources which are available from manuals supplied with each piece of chemical instrumentation. It is good practice to become familiar with an instrument by first reading the "Operation" manual. Most manuals will include a block diagram which identifies the key components in an instrument and shows the relationship between them. Usually a schematic diagram is also included. The schematic is used primarily by servicing personnel and shows exactly how each electrical component functions.

The block diagram can be used as a road map to the schematic which identifies specific components such as fuses, lights, tubes and terminals. It frequently provides answers to simple problems which may save hours or weeks of valuable research time.

Failures in the electrical section of an instrument are often caused by a fuse or a single tube. These are frequently replaced by the person who normally uses the instrument. An electronics specialist may provide advice about a problem, but the sole resource for getting exact information about the type of fuse or tube is the schematic furnished by the manufacturer. Look at Figure 27 and identify key features on the schematic such as transistors, fuses (type and kind), vacuum tubes (what kind, how many), etc.

The spectrometer uses a 0.5 A fuse, a 12AU7 vacuum tube in the amplifier circuit and an R106 photomultiplier tube in the detector circuit. Obviously, these numbers tell the electronics specialist a great deal about the characteristics of the component in question. However, the designations for components should be accepted only as model numbers. In another case, find the designation on the schematic for connecting the output to an external strip chart recorder. In actual practice you should look at the schematics for both the recorder and the spectrometer and determine the positive and negative (ground) terminals for both the input and output points. These are the kinds of things that are easy to find in a schematic and will be useful for putting an instrument into operation or troubleshooting problems.

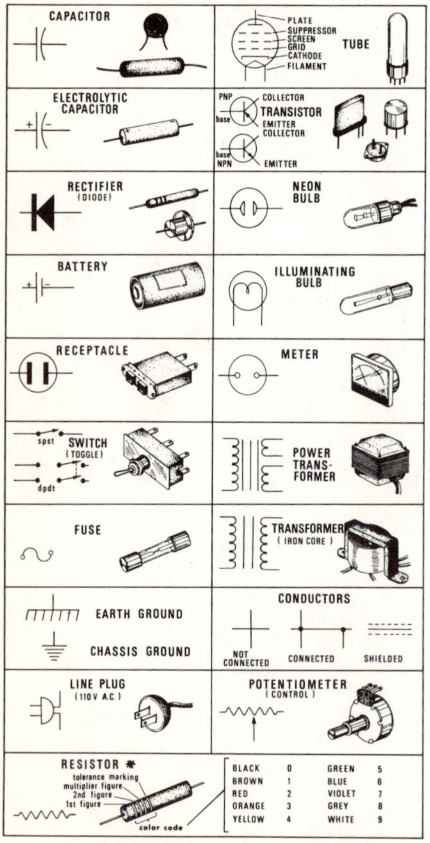

Figure 27. Standard electrical symbols

✱ Designations of Resistance Value

Most resistors are marked to show the nominal value of their resistance in ohms (a unit of electrical resistance). The present practice is either to print the value on the body of the resistor in numbers or to put three colored bands around the body of the resistor which can be translated by using the standard color code. Each number from 0 to 9 has been assigned a color as shown in Figure 27 The color of the band nearest to the end of the resistor represents the first figure of the resistance; the second band, the second figure; and the third band, the number of zeros to add to the first two figures to get the total resistance. Thus, a resistor with bands of yellow, purple and red would be 4700 ohms. Blue, gray, green would be 6,800,000 ohms or 6.8 megohms, and violet, green, black would be 75 ohms. For resistances between 1 and 10 ohms, gold is used for the third band. Thus, orange, white, gold is 3.9 ohms.

14

Using Laboratory Tools and Equipment

Now that you know how to translate an idea for a piece of apparatus into a drawing, you should understand how apparatus is actually built using various tools and materials. This involves knowing what materials are suitable for various situations and deciding which tools will help you most efficiently to construct your design. In this chapter the terms and tools used by specialists in machine work, glassblowing, and electronics are illustrated and described. Even if you never have to build anything yourself, your designs for equipment will be better and more practical if you understand and can avoid construction problems. This is only possible if you understand the techniques and tools involved in building laboratory set-ups.

Fundamental Machine Tools

The machine shop can produce parts made from metal or plastic stock material. Cylindrical surfaces are machined with a lathe. Flat surfaces are generally machined with milling machines or shapers. Holes are drilled with a drill press. Grinders are used for precision finishing of metal surfaces.

The Lathe. The primary function of the lathe is machining the surface of materials as they revolve. In the *facing* operation, a cutting tool is moved perpendicularly to the axis of the revolving object, removing materials and producing a plane surface (Figure 1). *Turning* is the process by which a cylindrical surface is cut to a smaller diameter (Figure 2). A cylindrical surface can be cut by tools having the profile of a thread space (Figure 3). This process, called *threading,* is fundamental in the design of metal and plastic materials which are closely coupled to give mechanical strength or a gas-tight seal while allowing them to be conveniently disassembled. Precision holes in materials are finished on a lathe in a process called *boring* (Figure 4). A tool is held in a boring bar on the lathe and moved parallel to the axis of rotation of the materials being formed.

The Drill Press. The precise drilling and countersinking of holes are done in a drill press. With a lathe, the material being worked turns, and the tools remain stationary. For the drill press, work is firmly clamped to the drill press table. The rotating drill, which is vertically aligned with the table, is brought to the work (Figure 5).

Milling Machines and Shapers. Frequently flat surfaces must be produced. Vertical and horizontal milling machines or shapers are used to machine flat surfaces on metal and plastics as well as to

137

Figure 1. **Facing.** *The chuck holds the part to be cut and revolves it in the direction shown. The tool in the holder is held against the work surface by the lathe carriage, and the motion of the tool across the face is controlled by the cross slide.*

Figure 2. **Turning.** *The chuck (a) holds the part (b) and revolves it in the direction shown. The tool in the holder (c) is brought against the work surface by the cross slide, and the motion of the tool is controlled by the carriage.*

Figure 3. **Threading.** *The chuck holds the part and revolves it in the direction shown. The tool, which has the profile of the thread space in the holder (a), is brought into the work surface by the cross slide (b).*

Figure 4. **Boring.** *The chuck holds the part and revolves it in the direction shown. The boring bar with the tool in the holder (a) is brought against the work surface by the cross slide. The advance is controlled by the lathe carriage.*

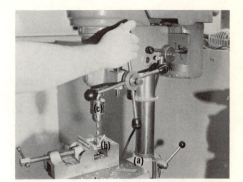

Figure 5. **Drilling.** *The table (a) supports the part (b), sometimes clamped as shown. The drill in the chuck (c) revolves and is forced downward by a hand-controlled gear in the drill press head to make the hole.*

Figure 6. Precision cutoff machines remove metal with hardened steel, carbide, or diamond cutting wheels. The work piece is held firmly and guided by the machine while the cutting wheel is brought into the work. Cutting produces high temperatures, so a spray of water is usually applied to the blade immediately after its contact with the work. Glass tubing is frequently cut on this machine by using a diamond wheel. Nonferrous metals and tubing are cut using abrasive wheels.

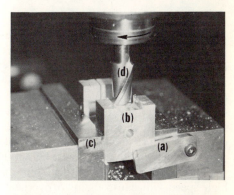

Figure 7. **Milling.** *The vice (a) holds the part (b) to the table (c), which moves laterally and longitudinally as shown. The milling cutter (d) revolves in the direction shown.*

Figure 8. **Grinding.** *The mandrel mounts and rotates the grinding wheel in the direction shown. The wheel moves laterally to traverse the entire surface to be finished.*

produce precise cuts on a surface (Figures 6 and 7). With a shaper the tool is held in a ram that moves back and forth across the work. A cut is made with each forward pass. Between cuts, the table supporting the work moves at right angles to the cutting tool so that closely spaced parallel cuts are made. Precise cuts (*e.g.,* grooves or slots) are generally made with a milling machine. The work piece is held firmly in a vice and moved under the rotating cutting tool. The rotating milling cutter may be positioned horizonally (horizontal mill) or vertically (vertical mill). Several cutters may be brought together in the horizontal mill to produce a number of parallel cuts at one time.

The Grinder. Grinding produces a smoother and more accurate surface than the turning or milling processes. An abrasive wheel turns at high speed and the work is brought into the wheel (Figure 8). Generally in precise finishing operations, a coolant flows over the work. Grinding wheels are chosen for their abrasive or cutting capacity and graded according to grit and composition of abrasive and binder materials. Grinding wheels are selected on the basis of the materials to be ground (*e.g.,* ferrous materials and aluminum would be finished on different types of grinders).

Hand Tools

A variety of small tools are used both in powered machines and as hand tools. Figure 9a shows a **twist drill** which is available in a number of sizes for producing holes in almost any material. Drills are rated by high speed or low speed types. Low speed twist drills are designed for use in hand drills or drill presses which do not turn faster than 600 rpm. Operating low speed drill bits at high speed will cause them to overheat and will ruin the bit. High speed bits can be used in equipment operating at all speeds. Power hand drills have the speed at which the drill turns stamped on the nameplate. The drill press is designed with a set of pulleys so that a variety of speeds can be obtained—from 720 to over 4,000 rpm. Drill bits and cutting speed are selected on the basis of the materials to be drilled and the size of the bit. Drill sizes are given as decimal numbers or fractions ranging from No. 40 .098 in. to .500 in.

A **countersink** is used to enlarge and alter the end of a hole to accommodate oval or flat screw heads. *Taps* are used for cutting the thread of a tapped hole. *Taper taps* are most commonly used for threading holes through metal or plastic. Plug or bottoming taps are used to finish cutting the thread into and through holes (see Figure 9e). If the thread must ex-

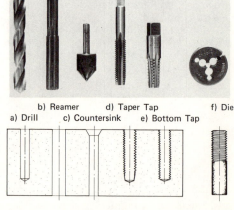

b) Reamer   d) Taper Tap   f) Die
a) Drill   c) Countersink   e) Bottom Tap

Figure 9.   Hand tools for drilling and cutting.

Figure 10. Tap handle and die holder.

tend to the bottom of the hole, generally the cut is started with a taper tap and finished with a bottom tap.

Taps must always be used with the properly designed tap handle (Figure 10). Never use makeshift handles such as pliers or wrenches. The correct tap handle allows the amount of pressure, torque, and alignment to be controlled during the threading operation; makeshift tools will result in weak or broken threads and broken taps. *Dies* (Figure 9f) are the complement to taps and are used for threading rods. For threading metals, cutting lubricants are used to insure a clean cut and their use is recommended.

There is a special terminology for screw threads. **Straight threads** have the same depth of cut throughout their length. This is the type of thread found on bolts, machine screws, etc. **Taper threads** are cut in a conical fashion so that the deepest thread is at the outside or end of a pipe or rod and it becomes more shallow as it gets farther from the end. Taper threads are used on pipe. The further a taper thread is screwed into

a fitting, the tighter it becomes. Right hand threads wind in a clockwise direction when viewed from the end at which the threads start. All threads are right hand (designated RH) unless specified otherwise. Left hand threads (designated LH) wind in a counterclockwise direction when viewed from the end at which they start.

Compressed gas cylinders use threaded fittings which may be either left hand or right hand (Figure 11). RH and LH threads are not compatible so a gas regulator with a left hand thread cannot be used on a gas cylinder with a right hand thread. This assures the manufacturer that the correct regulator will be used with the appropriate gas.

Threads per inch is the designation used to describe the form of threads. Coarse threads, designated as National Coarse (NC), are in general use and are used on most machine bolts. Fine threads, designated as National Fine (NF), are used in special applications. Threads for pipe are designated as National Pipe Standard (NPS). In desig-

Figure 11. A gas regulator thread.

nating threads, first the nominal size of the thread and then the number of threads per inch is given; *e.g.,* ¼–20 RH would be a standard right hand thread for a ¼ in. rod with 20 threads to the inch. Male threads are cut on rods. Female threads are cut in holes.

It is occasionally necessary to make accurate measurements of cylindrical objects. The tools used are the **outside caliper** and the **inside caliper** and an accurate scale (Figures 12–15).

A good method for setting an outside caliper to a scale is shown in Figure 12. The scale is held in the left hand and the caliper in the right hand. One leg of the caliper is held against the end of the scale (if it is a metal scale and has a true end); otherwise, start with the 1 in. or 25 mm graduation. The leg of the caliper is supported by the finger of your left hand while the adjustment is made with the thumb and first finger of the right hand. The proper application of the outside caliper when measuring the diameter of a cylinder is shown in Figure 13. The caliper is held exactly at right angles to the center line of the work and is pushed gently back and forth across the diameter of the cylinder to be measured. When the caliper is adjusted properly, it should easily slip over the cylinder due to its own weight. Never force a caliper or it will spring and the measurement will not be accurate.

To set an inside caliper for a definite dimension, place the end of the scale and the leg of the caliper against a flat surface. Hold the scale square with the flat surface. Adjust the other end of the caliper to the required dimension as shown in Figure 14. To measure an inside diameter, place the caliper in the hole as shown by the dotted line in Figure 15 and raise your hand slowly. Adjust the caliper until it will slip into the hole with a very slight drag. Be sure to

Figure 12. Setting an outside caliper.

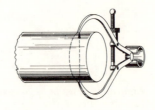

Figure 13. Using an outside caliper.

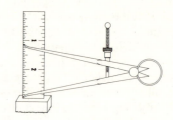

Figure 14. Setting an inside caliper.

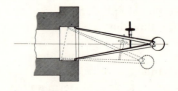

Figure 15. Measuring an inside diameter.

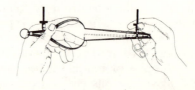

Figure 16. Transferring measurements from an outside caliper to an inside caliper.

hold the caliper square across the diameter of the hole.

In transferring measurements from an outside caliper to an inside caliper, follow the procedure shown in Figure 16. The point of one leg of the inside caliper rests on a similar point of the outside caliper. Using this contact point as a pivot, move the inside caliper as shown by the arrow in the illustration and adjust the thumb screw until you feel the measurement is just right.

Chassis punches are frequently called Greenlee punches. The punches are used for cutting large diameter, perfectly round, or square holes in thin metal. The die set consists of the three parts shown in Figure 17. In practice, a cres-

Figure 17. A die set for cutting holes in thin metal.

cent wrench is used to tighten the bolt which pulls a cutter into the die to make a precisely formed hole. Note the correct use of a crescent wrench in the figure. Always tighten bolts with a crescent wrench positioned so that the handle points in the direction in which the bolt is being turned; otherwise the jaws may spring open under pressure, slip, and round or break off the corners of the bolt head.

Common Metals, Wood, and Plastics

Angle metal. As the name indicates, it is possible to buy metals such as iron or aluminum formed into a 90°, or right, angle. As a structural shape, angles resist bending or twisting and can be fabricated into many different types of supports. There are four dimensions needed in specifying a piece of angle metal: 1) the size of each of the two legs, which may not be equal (*e.g.*, a typical dimension might be $1'' \times 1\frac{1}{2}''$); 2) length; 3) thickness; and 4) for angle iron, the nature of the web (Figures 18 and 19).

Metal channels are also common structural forms from which useful supports or structures can be constructed. The dimensions needed to identify a piece of metal channel are: 1) depth (or width); 2) flange height; 3) thickness; and 4) length.

Pipe and tubing. Seamless metal, rubber, or plastic tubing and pipe are used extensively in chemical laboratories and pilot plant operations. A knowledge of the qualities of these materials is necessary to determine when and where a particular type should be used.

Steel Angle Sizes

Dimensions, inches	Dimensions, inches
1/2 x 1/2 x 1/8	2 x 1 1/4 x 3/16
7/8 x 7/8 x 1/8	2 x 1 1/2 x 3/16
1 x 1 x 3/16	2 x 2 x 3/16
1 1/4 x 1 1/4 x 3/16	2 x 2 x 3/8
1 1/2 x 1 1/2 x 1/8	2 1/2 x 1 1/2 x 1/4
1 1/2 x 1 1/2 x 5/16	2 1/2 x 2 x 5/16
1 3/4 x 1 1/4 x 3/16	2 1/2 x 2 1/2 x 1/4
1 3/4 x 1 3/4 x 1/4	2 1/2 x 2 1/2 x 1/2

Steel Channel Sizes

3/4 x 3/8 x 1/8	1 1/2 x 3/4 x 1/8
1 x 3/8 x 1/8	2 x 1/2 x 1/8
1 x 1/2 x 1/8	2 x 1 x 3/16
1 1/4 x 1/2 x 1/8	2 1/2 x 5/8 x 3/16

Figure 18. Metal form sizes.

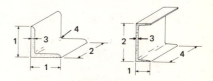

Figure 19. Dimensioning for angles and channels.

Pipes and Tubing

Metal pipe is usually made of steel or wrought iron. The size is designated by its inside diameter (i.d.). Galvanized pipe has a coating of zinc, inside and out, which prevents rusting. This factor must be considered when specifying a fabrication with galvanized pipe. Welding and cutting will frequently destroy the galvanized coating. Black pipe has the identical physical dimensions of galvanized pipe except that it lacks the coating and lends itself to machining, welding, and painting. Frequently it is used to make equipment or table supports. Black pipe is an approved material for transporting natural gas and acetylene (pipes with copper or brass may lead to the formation of highly explosive copper acetylide and are never used when acetylene or heating gases might be involved). Metal pipes are sized with the inside diameter of the pipe given as in Table I. Standard lengths are 21 ft. Brass, copper, stainless steel, and aluminum pipe have the same inside diameter as galvanized or black-iron pipe but they have thinner walls.

Seamless metal tubing is also available. It differs from the rigid pipe and is sold in rolls of up to 50 ft. Connections are made with machined fittings, and stand-

Table I. Common Copper Tubing Sizes

Soft (M, O), *sold in 50' coils* *i.d. (inches)*	*Hard (K, L),* *sold in 20' lengths* *i.d. (inches)*
1/8, 3/16, 1/4	1/4, 3/8
5/16, 3/8, 1/2	1/2, 5/8
5/8, 3/4, 7/8	3/4

ard sizes go from 1/16 in. inside diameter up to 1 in. Copper tubing is sold in four classes: K—extra heavy hard; L—heavy hard, M—standard hard, and O—light hard. M and O grades are commonly called soft tubing.

Plastic pipes are becoming more popular for handling liquids. PVC (polyvinyl chloride), hard rubber, and polyethylene pipes are available. Plastic pipes do not support combustion, are nonmagnetic, light weight, nonreactive in many systems, and can be fabricated by solvent cementing.

Pipes are connected using methods that depend upon the material and system. Welding, soldering, and threading are the most common methods. Figure 20 illustrates some common iron pipe fittings and shows the details of pipe threading. Figure 21 shows some common fittings used on seamless copper tubing. Figure 22 illustrates some typical fittings used with rigid plastic and rubber pipe systems.

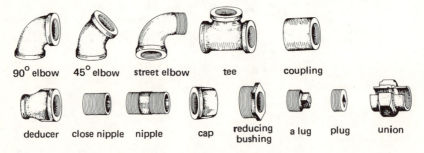

90° elbow 45° elbow street elbow tee coupling

deducer close nipple nipple cap reducing bushing a lug plug union

Figure 20. Typical iron pipe fittings.

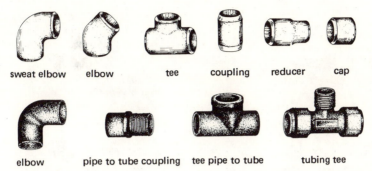

sweat elbow   elbow   tee   coupling   reducer   cap

elbow   pipe to tube coupling   tee pipe to tube   tubing tee

*Figure 21. (top) Typical copper pipe fittings. Figure 22. (bottom)
Typical plastic and rubber pipe fittings.*

Tubing Connectors

In general laboratory installations there are many other situations where plastic or metal tubing can be used and where quick, leak-proof reliable connections are important. Pressure, temperature, vibration, and surges have little or no effect on specially designed and machined tube fittings. Gas, liquid, and vacuum lines can usually be handled best with metal tubing systems. Typical manufacturers of tube fittings are identified by trade names such as Swagelok, Parker, and Imperial Eastman.

The chemical technician frequently will have to bend and connect soft metal tubing within apparatus setups. Connections may be made, for example, between a compressed gas cylinder regulator and a gas chromatograph using small diameter copper tubing. Glass systems in a fixed setup could be connected to a city water line using metal tubing with compression fittings.

Tube fittings are easy to install. For example, there is only one design of Swagelok fittings, regardless of material, size, or applications. Thus, there is only one set of instructions for connecting materials using this fitting. Details for handling these fittings are shown in Figures 23–26. Similar types of instructions apply to the fittings of other manufacturers.

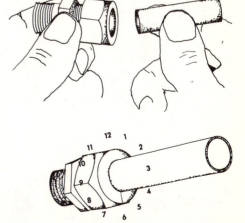

Figure 23. Simply insert the tubing into the tube fitting. Make sure that the tubing rests firmly on the shoulder of the fitting and that the nut is finger-tight.

Figure 24. Before tightening the nut, scribe the nut at the 6:00 o'clock position.

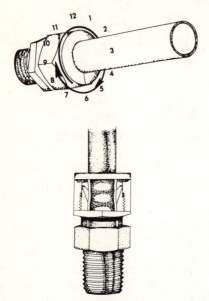

Figure 25. Now, while holding the fitting body steady with a backup wrench, tighten the nut one and one quarter turns (3/4 turn for 1/16 in., 1/8 in., 3/16 in. size tubes, or large tubing). Watch the scribe mark make one complete revolution and continue to the 9:00 o'clock position.

Figure 26. Fittings such as these provide a leak-proof, torque-free seal at all tubing connections and eliminate costly, hazardous leaks in instrumentation and process tubing lines. The tubing is supported rigidly over the entire length of the fitting. This provides leak-proof seals at three separate points. Two ferrules and a threaded chuck grasp tightly around the tube with no damage to the tube wall. There is virtually no constriction of the inner wall, thus insuring minimum turbulence.

When Swagelok-type fittings are to be used on equipment or apparatus where there are cramped conditions or where the body of the fitting is fastened to a light or fragile material, it is advantageous to use a preswaging tool on the tubing. The tubing, when preswaged with the nut and preswaged ferrules (Figure 27), can then be attached to a fitting by merely following the retightening instructions (Figures 28–31).

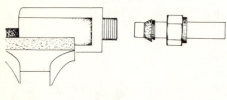

Figure 27a. Preswaging, Step 1. Using a sturdy support, a preswaging tool is used to swage ferrules onto the tubing to be installed by pulling the nut up one and one quarter turns from a finger-tight position.

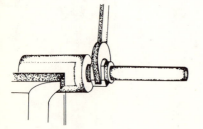

Figure 27b. Preswaging, Step 2. The nut is loosened and the tubing with preswaged ferrules is removed from the preswaging tool.

Figure 28a. Retightening, Step 1. Fitting shown in disconnected position.

Figure 28b. Retightening, Step 2. Tubing with preswaged ferrules inserted into the fitting until front ferrule seats in fitting.

Figure 28c. Retightening, Step 3. Tighten nut by hand. Rotate nut about one-quarter turn with wrench (or to original one-and-one-quarter tight position) then snug slightly with wrench.

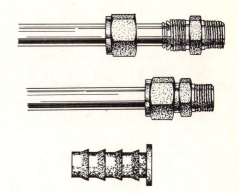

Figure 29. Insert for use with soft tubing such as Tygon and other soft types of poly-vinyl-chloride plastic tubing.

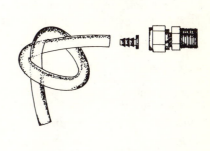

Figure 30. To determine the correct size of insert to be used, check outside and inside diameters of the plastic tubing.

Figure 31. This fitting provides a leak-tight seal on polyethylene tubing of all wall thicknesses without the use of inserts. The fitting is connected finger-tight to make a leak-proof connection on polyethylene tubing for vacuum or pressure applications. It consists of a brass tube fitting body, Zytel ferrules, and a brass knurled nut.

Table II. Plastic Fittings for Tubing

Tubing Material	Body	Front Ferrule	Back Ferrule
	Teflon[a]	Teflon	Nylon if temp. below 200°F. 316 SS if above 200°F.
Glass	Polyethylene[a]	Polyethylene	Nylon
	Nylon	Teflon	Nylon
	Metal	Teflon	Nylon or Metal (*see* above)
Teflon	Teflon[a]	Teflon	316 SS
	Nylon	Nylon	Nylon
Polyethylene	Polyethylene[a]	Polyethylene	Nylon
	Metal	Metal	Metal
	Nylon	Metal or Nylon	Metal
Nylon	Metal	Metal	Metal
Plasticized	Nylon	Nylon	Nylon
PVC or	Polyethylene[a]	Polyethylene	Polyethylene
Tygon[b]	Metal	Metal	Metal

[a]For more permanent type connections than usual laboratory work, it is found that a tighter grip on tubing can be obtained by using a metal nut on Teflon and polyethylene fittings.
[b]Brass, 316 SS or Nylon inserts must be used with these very soft types of tubing.

The availability of plastic fittings and ferrules as shown in Table II allows more applications, such as connection to glass. A variety of combinations of tubing and fitting materials is given to allow proper ferrule selection to obtain the best results and corrosion compatibility.

Compression tube fittings are used in much the same manner as the Swagelok-type fittings. A metal sleeve is inserted on the tubing and a slip nut compresses the sleeve into position on the tube

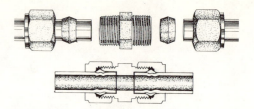

Figure 32. A compression tube fitting

within the body of the fitting (Figure 32). The sleeve and slip nut are supplied separately in this system, although they remain with the tubing once an assembly has been tightened.

Metal Tubing and Fittings

Before attaching fittings to tubings, it is necessary to prepare the tubing by: uncoiling, cutting, deburring, and bending.

To uncoil metal tubing, take the end of the tubing and, with one hand, hold it down on a flat surface such as a bench top or floor (Figure 33). Roll the coil away from the end of the tubing. Slide your hand along the tubing following the coil so the tubing lies flat. Do not uncoil

more tubing than is needed since repeated uncoiling and recoiling will distort and stiffen the tubing.

To cut tubing, use a metal tubing cutter (Figure 34). When cutting off a long piece of tubing, rock the tubing cutter with your hand, first going above the tubing, then back around, so that your hand goes below the tubing (Figure 35). This will allow you to cut the tubing without taking your hand off the cutter. The cutter handwheel can be adjusted as the rocking proceeds to maintain even tension on the cutting wheel. When cut-

Figure 33. Uncoiling tubing

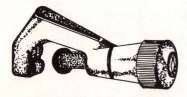

Figure 34. A tubing cutter

a) Cutting Long Tubing Pieces

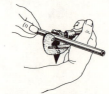

b) Cutting Short Tubing Pieces

Figure 35. Cutting tubing

Figure 36. Hacksaw and guide blocks used to cut tubing

a) Tube Cutters Burrs—
Inside

b) Hacksaw Burrs—
Inside and Outside

c) Removal of Inside Burrs with File

Figure 37. Burrs and their removal

ting a short piece of tubing, the cutter may be continually rotated around the end of the tubing. The handwheel can be gradually rotated to maintain even tension on the cutting wheel. **Do not use excessive pressure on the cutting wheel** or the tubing will be flattened. It is important that the cutting wheel be sharp and free of nicks.

If a tube cutter of the proper size is not available, a hacksaw can be used. Tubing should always be cut to length with a square cut. When using a hack-

saw to cut tubing, use guide blocks (Figure 36) to insure a square cut and to keep the tubing from flattening out. Do not use undue pressure when cutting soft tubing as the tube may flatten and thus be useless. While the cutters throw a burr into the inside of the tubing, the hacksaw will burr both the inside and outside of the tube (Figure 37a and b). Scrapers such as those attached to the handles of some cutters or files will clear the inside of the tubing (Figure 37c); the flat side of a file or emery cloth will remove the outside burrs.

Bends in tubing can be accomplished by hand or by using a bending device. Unless proper procedures are used, kinking or flattening will occur. This will weaken the tubing, causing a possible constriction or break (Figures 38–40).

The steps for making a bend with a hand bender are shown in Figures 41–44.

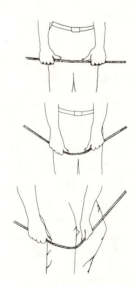

Figure 38. (top) Hand bending is accomplished by placing the thumbs towards each other but far enough apart to allow length for making the bend. Figure 39. (middle) Bend the tubing slightly. Figure 40. (bottom) Complete the bend.

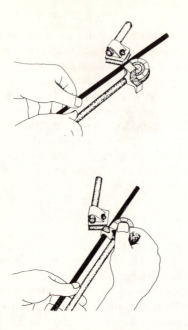

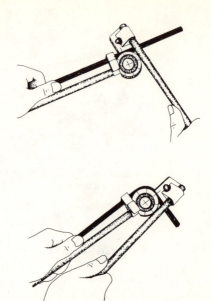

Figure 43. (top) The bending handle is brought into place. Figure 44. (bottom) The handles are brought together to bend the tube and bending is stopped when the desired angle of bend is shown on the angle indicator.

Figure 41. (top) The tubing is inserted in the bender. Figure 42. (bottom) The clamp is put over the tube.

Valves

Generally the design of any system involving gas, liquid, or vacuum lines should include adequate shut-off and control valves to maintain precise control. Shut-off valves use a plastic diaphragm disc, or metal plug which can be moved against a metal seat to assure a positive, leak-proof seal. Needle valves are designed to permit critical adjustments to the flow rate within a system (Figure 45). Several misunderstandings concerning the use of these types of valves should be mentioned: 1. needle valves are neither efficient nor effective within a system as shut-off devices— they are metering devices; 2. shut-off valves are not efficient as metering devices; 3. neither type of valve can control the pressure within a system; and 4. valves can introduce many foreign components into a system as a result of their construction. Thus, a well-de-

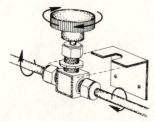

A valve in a line should be securely fastened to the rack or other solid fixture to avoid a strain and possible break in the tubing or fittings.

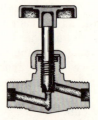

Figure 45. Fine threads on the needle valve stem permit critical settings to be made easily.

signed system involving lines where vacuum, gases, or liquids are controlled should have: 1. a shut-off valve; 2. a needle valve if flow control is important; 3. a provision to handle the maximum possible pressure or vacuum, such as a pressure regulator or an overpressure safety release; and 4. be constructed of the proper nonreactive materials, such as brass or stainless steel (Figures 46 and 47).

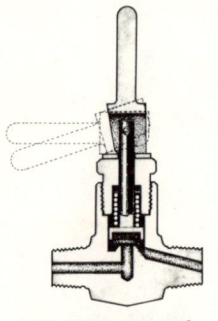

Figure 46. A lever-operated valve is designed for general operations where quick opening and closing is necessary.

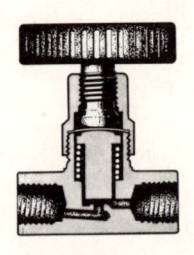

Figure 47. Metal diaphragm valves provide shut-off capabilities and a permanent leak-proof seal for the most critical conditions.

Machine Screws and Bolts

Selecting proper fastening devices is required for many laboratory setups.

The following figures (Figures 48–52) describe the more common fasteners, but there are many special ones for fastening problems.

Figure 48. **Bolts.** *A bolt is usually made of steel, brass, or stainless steel. It has a head on one end and a thread on the other end. Bolts are generally used with a nut to draw materials together when the nut is screwed on the threaded end.*

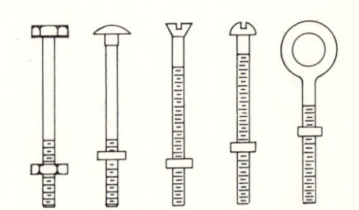

Figure 49. **Machine Screws.** *A machine screw is a small fastener used with a nut that functions in the same manner as a bolt. It can be used without a nut to work with a tapped hole. Machine screws are commonly made of steel, brass, or stainless steel. Typical sizes and styles of machine screws are shown in Table III.*

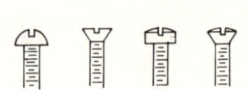

Figure 50. **Lock Washers.** *Many different styles of lock washers are used to prevent nuts from working loose. In addition to spring washers which are made of steel (hence may rust), there are patented washers with rust resistant coatings, distorted edges, or plastic inserts in common use as locking devices.*

Table III. Typical Sizes of Slotted Head Machine Screws

Designation	Diameter, in.	Threads per inch	Length range, in. (by 1/8 in. increments)
2—56	.086	56	1/8—1
3—48	.099	48	1/8—1 1/4
4—40	.112	40	1/8—1 1/2
5—40	.125	40	1/8—2 1/4
6—32	.138	32	1/8—3 1/2
8—32	.164	32	1/8—5
10—24	.190	24	3/16—6
10—32	.190	32	3/16—6
12—24	.216	24	1/4—6
1/4—20	1/4	20	1/4—6
5/16—18	5/16	18	3/8—6
3/8—16	3/8	16	3/8—6

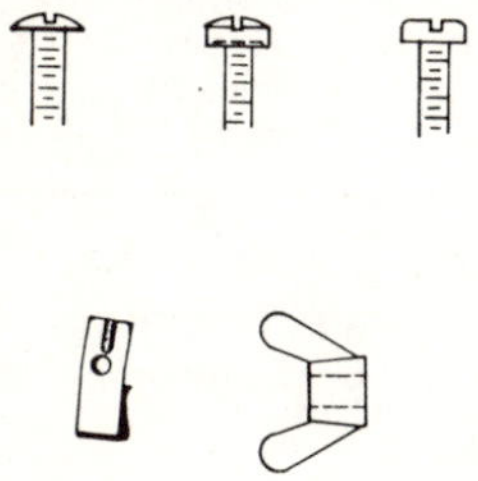

Figure 51. **Metal Screws.** *Self-tapping screws are made of hardened steel and form their own threads when driven into a hole of the proper size. For certain materials (e.g., sheet metal) these screws combine speed and reliability which makes them preferable over screw-nut combinations. Speed nuts used in combination with self-tapping screws give additional strength. See Table IV.*

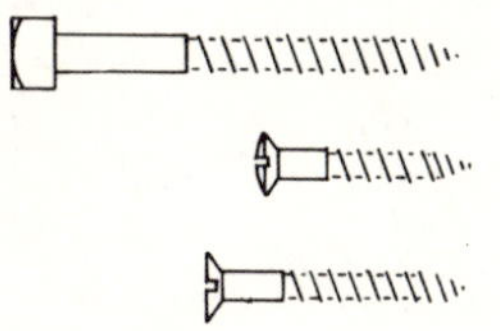

Figure 52. **Wood Screws.** *Wood screws are available in many forms and are made from a number of metals. They are sized by screw diameter and length. Heads may be slotted, have a Phillips-type head, or require a wrench (lag bolt). Several types of finishes are available (e.g., bright steel, nickel-plated, brass). They are usually made of steel or brass.*

Table IV. Typical Screws

Wood Screws			Self-Tapping Sheet Metal Screws		
Nominal size	Basic diameter, in.	Length range, in.	Nominal size	Basic diameter, in.	Length range, in.
2	.086	1/4—3/4	4	.112	1/4—3/4
3	.099	1/4—3/4	6	.138	1/4—1
4	.112	1/4—1 1/2	7	.151	3/8—1
5	.125	1/4—2 1/2	8	.151	3/8—1 1/2
6	.138	3/8—2 3/4	10	.177	3/8—2
8	.151	3/8—4	12	.216	1/2—2
10	.177	1/2—4			
18	.294	3/4—4			

Wood

There are three classes of wood-based materials which will be of interest to laboratory workers: finished lumber, plywood, and special materials such as Masonite or chipboard. The important thing to know about these materials is that the system of identifying dimensions for wood products varies. For example, finished lumber (pine or fir board, for example) is ordered by dimensions which do not reflect the wood wasted in finishing the product. For example, a board described as a 2 × 4 and 8 ft long would actually be 1⅝ in. thick, 3⅝ in. wide and 8 ft long. The thickness and width of finished lumber, the kind commonly bought and sold, is ⅜ in. less than the dimensions requested. The length would be exact. Table V lists common sizes of wood.

Plywood, Masonite, and chipboard are sold in sheets exactly 4 ft wide and 8 ft long. Common thicknesses for plywood are ⅛ in., ¼ in., ⅜ in., ½ in., ⅝ in., and ¾ in. Plywood is sold in exterior (made with moisture resistant glue) or interior (not resistant to moisture) types. It is graded such as "good on one side" (free from holes or plugs), or "good on both sides," or "construction grade" (generally a semi-rough surface with knots replaced with wood plugs).

Table V. Typical Softwood Description

	Grades
Clear:	kiln dried, knot free, vertical grain
No. 1 or construction:	tight knots, and free of knots on edge of board
No. 2 or standard:	large knots and possibly knots on edge of board
Utility:	open knot holes

Standard sizes of dimension lumber (in.)
1 x 2, 1 x 4, 1 x 6, 1 x 8, 1 x 10, 1 x 12
2 x 4, 2 x 6, 2 x 8, 2 x 10, 2 x 12, 4 x 4

Standard board lengths (ft)
6, 8, 10, 12, 14, 16, 18, 20

Rubber and Plastic Flexible Tubing

Rubber or plastic tubing is used in almost every laboratory setup the chemical technician will encounter. Various types of tubing have special characteristics which should be understood. Black rubber seamless tubing is made from a pure gum rubber of a specially selected grade and with a low sulfur content. It is used commonly for transferring water, vacuum, and solutions which would not react with the rubber. Medium wall rubber tubing is used where pressure or vacuum is not a factor. Heavy wall rubber tubing is used for vacuum. *See* Table VI.

Tygon tubing is flexible and clear. It can be steam autoclaved or chemically sterilized. A copolymer formulation, Tygon, is advertised as safe to use with virtually any chemical. This feature can be misinterpreted. Any installation in which concentrated acids, bases, or solvents are transferred through plastic or rubber tubing should be checked to insure that no reaction will occur or that contaminants, such as the plasticizer in the tubing, are not introduced into the chemical system. Polyethylene, vinyl, and polyvinyl tubing are also available. The chemical and physical characteristics of these materials are given in Table VII.

Table VI. Typical Rubber Tubing Descriptions

RUBBER TUBING, Black, Hand Made—Cloth impressed surface and medium wall. Pure gum tubing; seamless, with low sulfur content and very durable for general laboratory purposes. (Boxed in 50 foot lengths).

RUBBER TUBING, Amber Latex— It is translucent and is recommended for all kinds of glass connections in which a high quality clinging tubing, highly elastic, and of long life, is desired. Special packing on reels in dispenser boxes except sizes 1/4 x 1/8", 3/8 x 3/32", 1/2 x 1/16" and 1/2 x 1/8".

id, inch	Thickness of wall, in.	id, in.	Thickness of wall, in.	Feet to reel or box
1/8	1/16	1/8	1/32	100
3/16	1/16	3/16	1/16	100
1/4	1/16	3/16	3/32	50
1/4	1/8	1/4	1/16	50
5/16	1/16	1/4	1/8	50 (box)
3/8	1/8	5/16	1/16	50
1/2	1/8	5/16	3/32	50
3/4	1/8	3/8	3/32	50 (box)
1	1/8	1/2	1/16	50 (box)
		1/2	1/8	25 (box)

Table VII. Tygon Tubing Description

Inside diameter, in.	Wall thickness, in.	Outside diameter, in.	Package, ft
1/16	1/32, 1/16	1/8, 3/16	50, 100, 2000
1/8	1/32, 1/16	3/16, 1/4	50, 100, 500
3/16	1/32, 3/32	1/4, 3/8	50, 100, 500
1/4	1/32, 1/16, 3/32, 1/8	5/16, 3/8, 7/16, 1/2	50, 100
4	1/2	5	50

Extra heavy wall for use as vacuum tubing

3/16	3/16	9/16	
1/4	3/16	5/8	
3/8	1/4	7/8	
1/2	5/16	1 1/8	
3/4	3/8	1 1/2	

To order tubing, specify desired i.d., o.d., wall thickness, and length.

Welding Metals

The subject of permanently fastening things together is of particular interest to the chemical technician since plastic, metal, and glass components are constantly being fabricated or designed. Understanding the limitations or advantages of various bonding techniques, unique in most cases to the materials being handled, is vital.

Welding processes are classified according to how the welded joint is completed. In arc welding, pieces of metal to be welded are brought to the proper welding temperature at the point of contact between an electrode and the metal. The heat liberated by the .electric arc melts the metal electrode and the pieces of metal to be joined so that the metals are completely fused into each other.

Extraordinarily strong welds can be produced by electric arc welding. This is a common technique for joining steel pieces.

In gas or oxyacetylene welding, a high temperature flame is produced by igniting a mixture of two gases, usually oxygen and acetylene. A welding rod is melted in the flame as the metals to be joined are brought up to fusion temperature. The molten rod material flows into the joint between the metals. This technique is used for joining steel and iron materials.

Brazing is a variation on gas welding. Brass welding rods and a special flux are used to join nonferrous materials such as brass or copper.

Inert gas, or Heliarc, welding is used to join aluminum metals together. An electric arc is used to bring the metals to fusion temperature. An aluminum electrode is used. A special electrode holder forces helium gas around the electrode and molten metals to form an inert gas envelope. This process prevents the formation of aluminum oxides which weaken conventional welds.

In certain cases it is not necessary to bring the metals to be joined to their melting temperature. The techniques of silver-soldering are frequently used to fabricate metal parts at relatively low temperatures.

Silver-solder is an alloy of copper, zinc, and silver which melts at approximately 1400°F. Silver-solder requires a flame similar to brazing. For metal parts which have closely machined tolerances that must not be changed and in which heat could warp or damage the machined areas, silver-soldering is a good choice.

Solder is commonly a mixture of tin and lead. It melts at approximately 600°F. Some solders have a hollow core containing a flux in resin form. Open flame torches are not required for soldering. Electric soldering irons are most commonly used. For joining light pieces of steel (sheet metal), solder with an acid flux makes strong water-proof joints. Electric wires are usually copper, and solder with a resin flux (not acid) is the most commonly used means of joining them together.

Laboratory Glassware

Most individual laboratory work is done with glass vessels or apparatus. Most glassware is purchased from laboratory supply houses, and for routine requests or orders it is only necessary to identify the names and categories of standard glassware items. However, new or modified laboratory procedures may require that special apparatus be made, modified, and maintained. It is very useful to be familiar with the possibilities and limitations of glass and be able to discuss apparatus design with other laboratory workers or professional glassblowers.

The history of glass manufacturing goes back about three thousand years. Many glass mixtures now exist, but most laboratory workers usually encounter only a few of them. (There are many special types.) It is important to be able to distinguish among the types of glass commercially available since they differ greatly in resistance to attack by chemicals, physical features (such as resistance to physical shock or temperature differences), and cost.

Most glass materials used in the chemical laboratory are made from **borosilicate glasses.** These glasses have a low coefficient of thermal expansion and are less likely to break when sud-

> *Coefficient of thermal expansion: the in-crease in length per unit length per degree rise in temperature.*
>
> Pyrex No. 7740 glass expands 32.5 parts in 10 million for each °C rise. Soft glass ex-pands 92 parts in 10 million for each °C rise. Copper metal expands 167 parts in 10 million for each °C rise. Cooling joined materials of different coefficients of expan-sion causes them to pull apart, break, or become severely strained.

denly heated or cooled. They also have good mechanical strength. Borosilicate glasses are resistant to chemical attack which reduces contamination of systems by materials leached from the glass. The Corning Glass Works' borosilicate glass No. 7740 has been named Pyrex (Table VIII). Mistakenly, many labora-tory workers apply the Pyrex brand name to all glassware and apparatus made from borosilicate glasses. Kimax

Table VIII. Technical Data from Corning Glass Works

Glass No. 7740 is a low alkali-content, borosilicate glass. It is free from mag-nesia-lime-zinc group elements, heavy metals, arsenic, and antimony.

Thermal Durability
    Strain point— ∼515°C.
    Annealing point— ∼565°C.
    Softening point— ∼820°C.

In general, a glass article will not deform when heated to its Annealing Point. However, if it is subjected to prolonged heating or pressure, the highest safe operating temperature is its Strain Point. If properly supported, and not under high internal pressures, some pieces of labware made of No. 7740 can be heated above 600°C for a relatively short period of time. However, any glass heated above its Strain Point and then quickly cooled will acquire strains which may affect future serviceability. You can reduce the possibility of strain by cooling the glass article slowly and uniformly.

glassware produced by the Kimble Glass Company has a similar composition to Pyrex.

Lime, soda, or **flint** glass are names for older and comparatively inexpensive formulations. Heating expands lime glasses more than borosilicate glasses. Thus, rapid temperature changes may damage apparatus made from lime glass. Also, lime glasses are not as chemically resistant as the borosilicate glasses. One advantage of lime glass is that it can be readily worked in a com-pressed-air gas flame since the softening temperature of this glass is only about 560°C. It is very difficult to make com-plex glassware from lime glass since special care is needed for heating and annealing it. Kimble glassware type No. R-6 is an example of commercially-available laboratory glassware made from lime glass (Table IX).

If you are to work with light-sensitive materials, you may wish to use **actinic**

Table IX. Technical Data from Kimble Glass Co.

Standard Flint (R-6) Glass is a superior soda-lime glass which has been available for many years. It was developed pri-marily for vessels which will be used at or near room temperature. In this temp-erature area, the chemical durability of Standard Flint Glass is such that negligible reaction will take place be-tween the glass and liquids. R-6 Glass meets the requirements for Type II glass of Federal Specification DD-G-54lb, "Glass, Laboratory." R-6 Glass is the glass of choice for weighing bottles and micro apparatus in general since the glass is less prone to develop static surface charges than borosilicates.

Thermal Durability—
    Strain point—490°C
    Annealing point—520°C
    Softening point—700°C

Table X. Light-Transmittance of Actinic Ware

Wave length (mμ)	Transmittance (%)
300	0
400	1
500	4

beakers or glassware. It has been specially coated to give a permanent red color that allows safe handling of materoals sensitive to light in the 300 to 500 mμ range (Table X).

For high temperature requirements glassware produced from fused silica or a 96% silica glass, such as Corning Vycor, is useful. Silica labware is recommended for use requiring very high temperatures, drastic heat shock, or extreme chemical resistance. It will withstand continuous temperatures up to 900°C and from that temperature can be placed in ice water without breakage (Table XI). This glassware is expensive and requires special fabrication tools.

Fritted glassware uses discs or cylinders of porous glass. It is made entirely from borosilicate glasses and is used for standard filtering devices. A glass frit can be cleaned in about the same way as standard glassware. Absorption tubes

Table XI. Resource Information for High Silica Glass

Vycor Brand Laboratory Glassware Glasses Nos. 7900 and 7913
Maximum recommended temperature for continuous use of blown ware is 900°C; for tubing and tubing articles, 1000°C.

Thermal Durability

	Glass No. 7900	Glass No. 7913
Expansion (x 10^{-7})	8	8
Strain point	820°C	890°C
Annealing point	910°C	1020°C
Softening point	1500°C	1530°C

made from fritted glass are used in systems involving gas washing, absorption, or extraction.

Most laboratory apparatus and glassware is made from borosilicate glass. Unless otherwise specified chemical storage containers should be assumed to have the characteristics of lime glass. Thus, you should be certain that glass containers are suitably resistant when storing concentrated alkali or other solutions which could attack lime glass.

Technicians occasionally have to order glass tubing and rod for their work. These products are supplied in four-foot lengths. The diameter of glass tubing, capillary tubing, and rod is specified in

Table XII. Catalog Resource Information for Fritted Ware

Designation	Catalog Abbreviation	Nominal Max. Pore Size (microns)
Extra-coarse	EC	170-220
Coarse	C	40-60
Medium	M	10-15
Fine	F	4-5.5
Very fine	VF	2-2.5
Ultra-fine	UF	0.9-1.4

Cleaning—A new fritted filter should be washed by suction with hot hydrochloric acid and then rinsed with water before it is used. Clean all fritted filters immediately after use. Many precipitates can be removed from the filter surface simply by rinsing from the reverse side with water under pressure not exceeding 15 lbs/sq in. Some precipitates tend to clog the fritted filter pores and chemical cleaning is required.

Material	Cleaning Solution
Fatty materials	Carbon tetrachloride
Organic matter	Hot, concentrated cleaning solution
Mercury residue	Hot nitric acid
Silver chloride	Ammonia or sodium hyposulfite

millimeters. Tubing is available in three wall thicknesses—standard, medium, and heavy. Glass tubing is sold by the pound. When ordering glass tubing or rod, specify the amount in pounds, the diameter in millimeters, the wall thickness, and the kind of glass desired.

Glass tubing should be stored flat and kept clean. The type of glass should be carefully marked on the container.

Glass of all types looks and feels very much alike; it is easy to confuse borosilicate and lime glasses on visual examination. Lime glasses cannot be successfully joined to the borosilicate glasses; nor can the borosilicate glasses be joined to the high sylica types such as Vycor. There are tests available to identify types of glass. First, the melting or softening points differ. Also, Pyrex No. 7740 glass can be distinguished from lime glass by immersing it in a solution of 16 parts by volume of methyl alcohol and 84 parts by volume of benzene. This solution has a refractive index that is the same as for Pyrex No. 7740 glass (1.474). If a questionable piece of glass seems to disappear when immersed in the solution, it is Pyrex No. 7740. Other types of tubing will remain plainly visible.

Special glasses are frequently useful for specific purposes, especially when extra strength or chemical resistance is required. Considerable information about the glass is given in manufacturers' catalogs.

Cutting and Heating Glass Tubing

Glass tubing, rods, and capillary tubing must often be cut to the desired length. For rods and tubing up to about 20 millimeters in diameter, the glass should be **scored** with a triangular file. To do this, simply hold the glass flat on a table with one hand and press down on the file and push it away from you (with the file handle pointing toward your body) to make a single notch perpendicular to the axis of the tubing. Only make one single, straight cut with a smooth movement of the hand (Figure 53). Once the scratch has been made, moisten the scratch with water. While protecting your hands with a towel, grasp the rod or tubing with a hand on each side of the cut. With your thumbs placed directly behind the scratch, pull your hands away from each other and push with your thumbs (Figure 54).

Tubing with an outside diameter greater than 8 mm may require several strokes, starting at the end of each mark, with a sharp file to score at least one-third to one-half of the circumference.

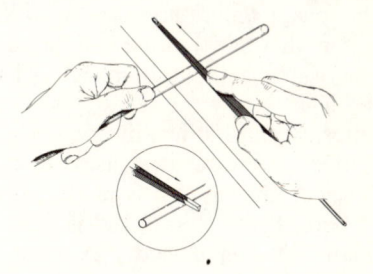

Figure 53. Scoring glass tubing

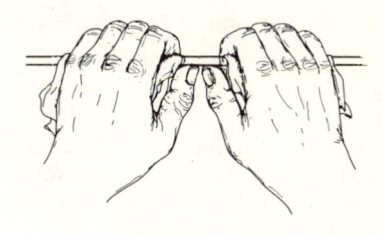

Figure 54. Breaking glass tubing

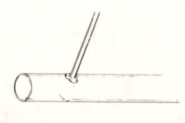

Figure 55. Use hot glass to break large tubing

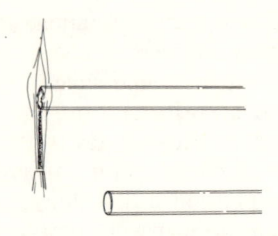

Figure 56. Firepolishing glass tubing

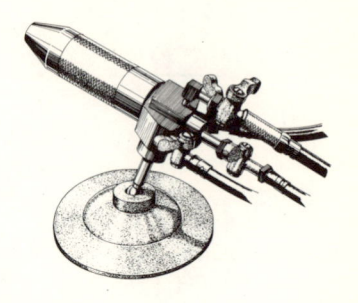

Figure 57. Bench burners for glass working

Tubing larger than 20 millimeters should be cut in a glass shop, although applying hot glass to the scratch frequently causes a break (Figure 55). Some references for glass working include:

1. R. Barbour, "Glassblowing for Laboratory Technicians," Pergamon, London

2. W. E. Barr and Victor J. Anhorn, "Scientific and Industrial Glass Blowing and Laboratory Techniques," Instruments, Pittsburgh, Pa.

3. J. E. Hammestzhr and C. L. Strong, "Creative Glassblowing," W. H. Freeman, San Francisco

4. E. L. Wheeler, "Scientific Glassblowing," Interscience, N. Y.

Unless the glass is to be immediately joined with other glassware, it should always be *firepolished* because the edges of freshly cut glass are very sharp. This is done by melting the edges in the fire (Figure 56).

A chemical technician should be able to make simple connections and perform certain manipulations with glassware. This requires a burner that can produce a hot enough flame to melt the glass. The burner must allow control of the shape of the flame for some work.

The flame of a glass working torch must have a sharp boundary in order to concentrate heat where it is wanted. Temperatures must be made rather high, and flames are produced by mixtures of either gas and compressed air or gas and oxygen. The mixture must be adjusted to meet specific requirements. An ex-

ample of a bench burner for glass working is given in Figure 57.

The gas hand torch is shown in Figure 58. A hand torch is necessary for connecting a stationary configuration of glass tubing such as a vacuum system and for simple apparatus that cannot or should not be taken to the glass shop. Figure 59 shows the mixture needed to produce

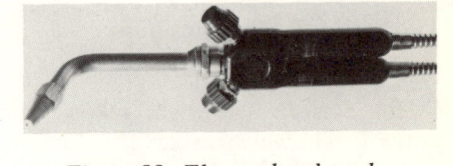

Figure 58. The gas hand torch

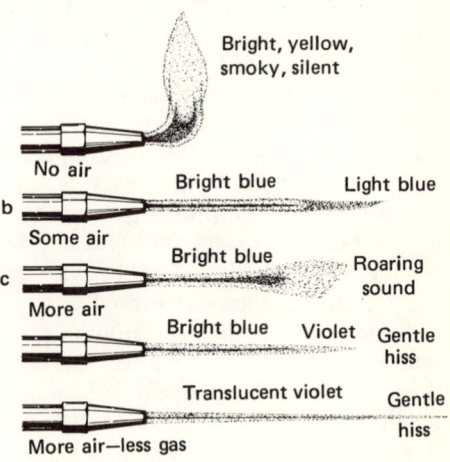

Figure 59. Flame adjustment

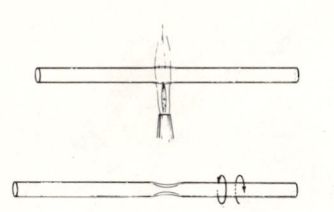

Figure 60. Heated glass section thickens naturally

an adequate glass working flame. A properly adjusted gas-oxygen flame produces a brilliant, bluish-green inner cone surrounded by an almost invisible outer cone of violet.

Before using a glass torch you need to know the characteristics of hot glass. As glass becomes soft, it responds to gravity. Manipulations should take advantage or offset the effects of gravity. This is usually accomplished by rotating the work. If a horizontal piece of glass tub-

ing is heated without rotation, softening and sagging will occur, and it becomes difficult or nearly impossible to restore it to its straight, original form.

Thin glass expands faster than thick glass. If a piece of thick glassware is joined to a piece which is thin, the abrupt application of heat will produce dissimilar expansion rates and cause cracking or breaking. Heating glass naturally thickens and pulls it together. Try heating a piece of glass tubing while rotating it with your fingers as shown in Figure 60. Capillaries or constrictions can be made by using this characteristic. Uniform thinning is produced by pulling the heated glass tubing. Although it is a useful technique for straightening joints, pulling must be used cautiously to avoid thinning the glass and producing sections of glass having different thermal characteristics.

Tools and Techniques for Working Glass

Hot glass gives a bright yellow color to the flame, which makes it difficult to see small details. When a worker is making small or intricate pieces, didymium glasses can be worn to block out the yellow light. Other basic tools that should be handy for glass working are asbestos paper or board on which to set hot glassware, a collection of stoppers to seal ends if blowing is necessary, a stainless steel gauge, sharp files, a flaring tool, a glass blower's swivel, and tweezers for pulling capillaries, adding hooks, and other manipulations (Figure 61).

Handling glass tubing requires practice. Glass should be heated no more than necessary to perform the desired manipulation. Beginners tend to overheat materials. Glass, when sufficiently plastic for proper manipulation, is still

rather rigid and gives one a feel for movement. The key to glass working is to heat the glass uniformly and rotate it to prevent sagging (Figure 62). A good way to improve skill for rotating glass is

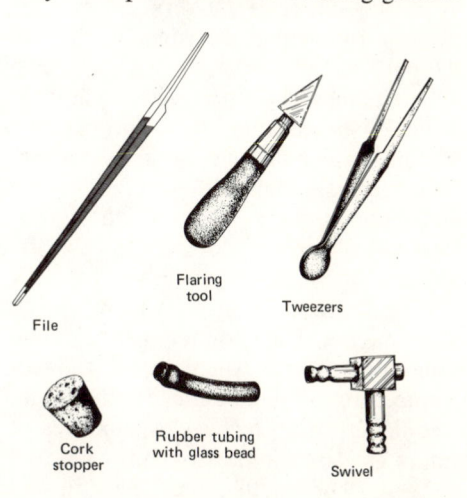

Figure 61. Tools for glass blowing

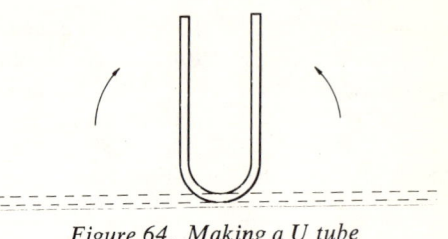

Figure 62. *Improper heating causes sagging*

Figure 64. *Making a U tube*

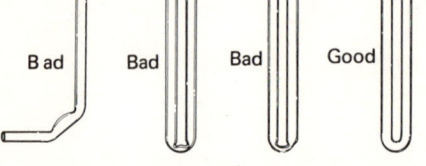

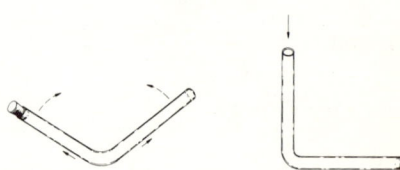

Figure 63. *Heating glass tubing for bending*

Figure 65. *Bending large diameter tubing*

to rotate two pencils, one in each hand, so that the eraser ends are always in alignment.

Next to cutting and fire polishing glass tubing correctly, the ability to make bends up to 180° is the most frequently used talent. To make a good bend, simply rotate the glass tube in a flame as shown in Figure 63 while simultaneously moving it back and forth to soften an area four times the tubing diameter (24 mm long for a 6 mm diameter tube). When the tube has become sufficiently plastic, remove it from the flame and lift the ends of the tube to form the desired angle (Figure 64).

Problems increase with tubing size when bends are to be made in tubing larger than 8 mm diameter. Bends of large radius or in large diameter tubing must be made with a combination of pulling, bending, and blowing (Figure

65). Try to avoid making long sweeping bends. This is a job for a glassblower with special burners. The figures show the result of some common mistakes (Figure 66).

Expanding the end of a tube into a conical or funnel shape is called *flaring* (Figure 67). The end of the tube is rotated and softened in the flame. The tubing is removed from the flame and a smooth metal rod, called a flaring tool, is placed about ⅜ in. into the opening and

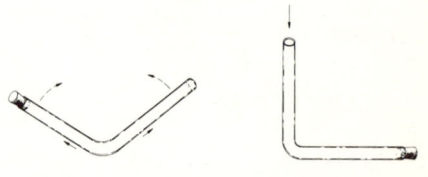

a

Heat

Rotate against flaring tool

b

Work flare outward

c

Expand into cone

d

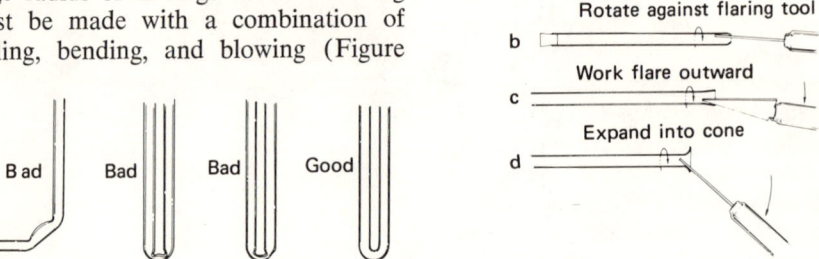

B ad   Bad   Bad   Good

Figure 66. *Comparing bending results*

Figure 67. *Flaring glass tubing*

gently pulled against the softened edge of the glass. The resulting force of the metal against the glass stretches the glass and expands the end of the tube into a cone. The edge of the glass can be pushed out a little at a time. The work will probably cool, become too stiff, and will have to be returned to the flame several times to be reheated to the working temperature. As the flare diameter increases, you should extend the length of the cone slightly.

Joining Tubing

Several steps are required to join glass tubes of the same size. Close one end of a tube with a stopper to facilitate blowing. The end of the other tube must be either firepolished if it is to be put into the mouth or connected to a rubber tube if the glass position will not allow free rotation. Position your hand and torch so that your position will not have to be changed when the glass reaches working temperature.

Next, adjust the torch as shown in Figure 68. Bring the tube ends into the flame and begin to melt them. When the tube ends are just melted so that the glass flows together when the ends are touched, press them together gently. The bore of the tubes must be exactly in line. Heat the joined area until the ends are well melted together. If possible, rotate the tubes for the entire time while heating.

When the ends have fused, remove the tube from the fire and, after about a second, blow gently to expand the glass so it is only slightly thicker than the nearby tube wall. Then, while continuing to rotate the tube, pull gently—just enough to straighten the seal (Figure 68).

It is good practice to anneal any glass which has been heated to its softening temperature. Internal stresses develop when the molten glass cools, but can be prevented by cooling the glass slowly so that all parts of the piece cool at the same rate. Partial annealing is done by slowly removing the glass from the flame after the work has been completed. A more effective way is to reduce the air supply to the torch and rotate the glass in the luminous flame until it is covered with soot.

Joining glass tubing with unequal diameters or different wall thicknesses requires manipulations that are somewhat different from those for sealing tubes of equal diameters. Joining tubes of un-

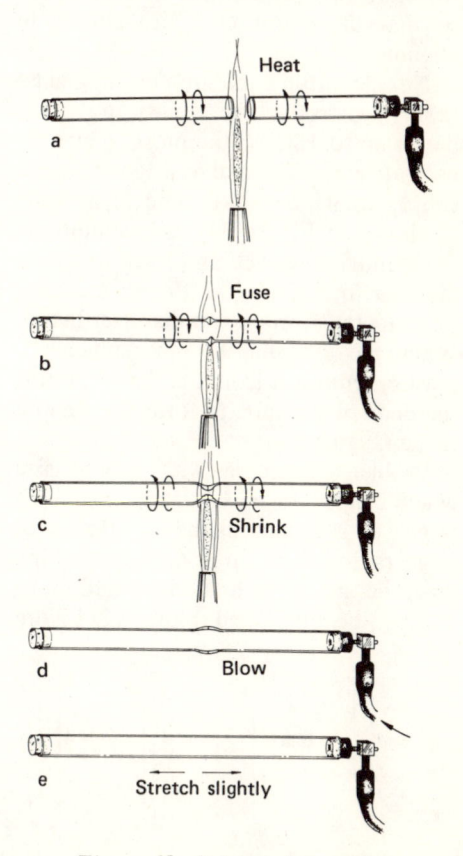

Figure 68. Joining glass tubing

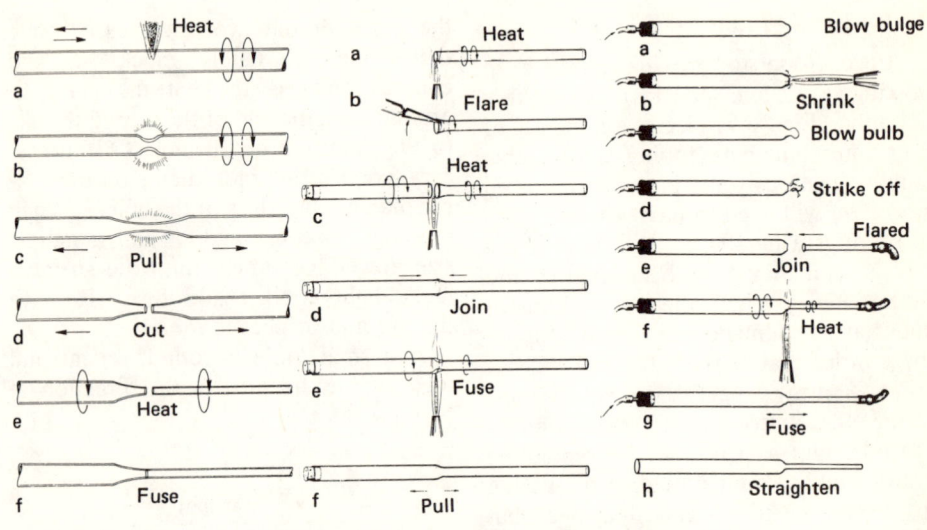

Figure 69. Joining different sized tubing

equal diameters can be done by either shrinking one end of the larger tube to match the smaller tube or by flaring the end of the smaller tube to match the larger tube. A combination of both methods might be needed (Figure 69).

Once an approximate fit has been obtained between the two pieces of tubing, proceed as if joining tubing of similar size as described in Figure 68.

Hose connections to glass apparatus can be firmly made by working knurls or tubulations on glass tubing. The knurls facilitate wiring tubing firmly in place.

To make knurls, concentrate the flame on a narrow part of the glass tube. Rotate the tubing and remove it from the flame as soon as the glass becomes slightly plastic. Push the ends of the tube together to make a low, encircling bulge. Continue heating and pushing until a ring has been formed which is about 1 mm higher than the normal tubing diameter. Repeat the process to form another knurl about 5 mm from the first knurl.

To facilitate sliding rubber or plastic tubing onto the glass knurls, it is useful to pull the glass to slightly reduce the outside diameter before cutting and fire-polishing. These steps are shown in Figure 70.

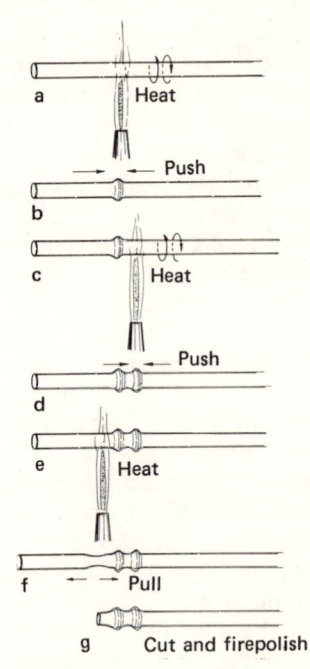

Figure 70. Putting knurls on tubing

Making a T–Tube

An exercise to provide practice in making knurls and seals involves making a T-tube (Figure 71). Cut two pieces of 6 or 8 mm lime glass tubing to a suitable length. There is some planning involved since you will need to have enough glass to serve as a handle on each end of the knurls until they have been cut off and firepolished. When the three knurled ends have been made and cooled, stopper the knurled end of the tube that will become the arm of the T. Also stopper one of the other knurled ends. Attach a blowing tube and a swivel to the remaining knurled end. Practice holding the tubing in several positions and find one that seems most convenient so you will not have to change hand positions when the glass is brought to a working temperature.

Light and adjust the burner flame to give a needle-point flame. Hold the tube to be used for the straight section of the T so the bright, inner cone of the flame is almost, but not quite, in contact with the center of the tubing. Within seconds a small area of glass will redden and sag slightly to form a shallow dimple. Let the dimple expand to a width of about ¼ in. Remove the flame and blow the dimple into a bulge about as high as it is wide. Then heat the top of the bulge with the needle-point flame until half of the area sags. Remove the flame and blow a bulb about ⅜ in. in diameter. Break off the bulb with the edge of a wire mesh screen and brush the fragments away. Take the tubing that will become the leg of the T and hold it vertically over the hole just made in the crossarm. The openings should be spaced about ⅛ in. apart.

Preheat both the end of the tubing and the edge of the hole with the flame and then direct the flame simultaneously on the edges of the two openings. Rotate the flame alternately clockwise and counterclockwise to heat the glass all around until the end of the tubing shrinks to the diameter of the hole. (If necessary, concentrate more flame on the tubing than on the edges of the hole.) When the edges become molten and match in size, lower the tubing until the surfaces make light contact and fuse. Remove the fire and inspect the joint.

If a small hole is found, reheat and then incline the tube in the direction of

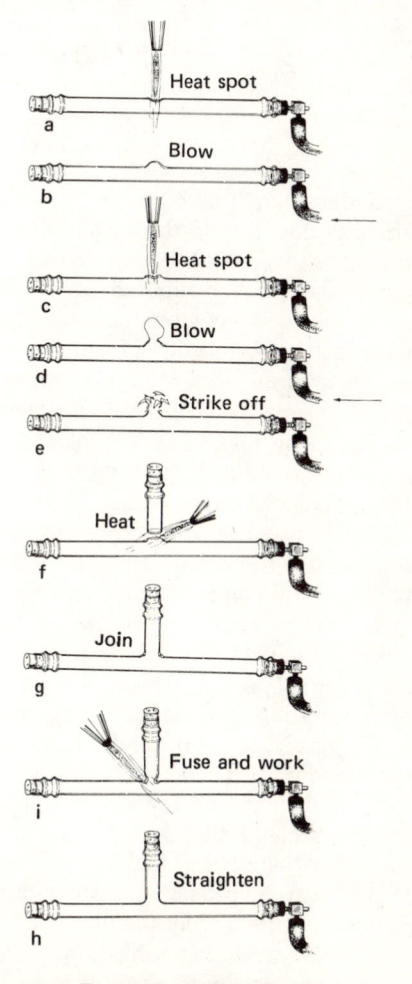

Figure 71. Making a T

the leak, wobbling it a bit if necessary to close the hole. Then return the tube to the vertical position and lift it a fraction of an inch to stretch the glass slightly. Remove the flame and immediately suck on the blowhose just enough to constrict the joint perceptibly. Failure of the joint to constrict under the partial vacuum may indicate the presence of a second, previously undetected, leak.

Close the leak by the technique just described. Apply suction again. When the joint constricts, indicating a good seal, blow the glass out to its former dimension and while holding the tube motionless, continue to heat the work until the junction shrinks about 20%. Remove the flame.

Let the material cool about 3 seconds. Then simultaneously stretch the glass and expand it by blowing until the bore and wall thickness of the seal match that of the tubing. Cool gradually by passing the flame up and down over the glass.

Graded Seals

Vacuum equipment is often constructed of non-borosilicate material which requires **graded seals** because it must be joined to a glass manifold. These seals are made from different glasses, starting with borosilicate glass, with each successive glass having a higher or lower coefficient of expansion and a higher or lower softening point. The series ends with a glass which fuses readily with soft glass or quartz. Graded seals are available from many glass companies. Examples of graded seals are:

1. **Pyrex-to-Soft Glass.** Many types of electronic equipment made of soft glass must be joined to borosilicate glass. The equipment can be attached by first joining it to the lime glass end of a Pyrex-to-lime glass seal. The other end can then be joined to a borosilicate glass.

2. **Pyrex-to-Vycor and Quartz.** The high-silica end of the seal can be fused to either Vycor (96% silica) or fused quartz (99.9% silica). If a quartz or Vycor apparatus is to be evacuated through a standard joint (a ground glass joint) connection, it is much less expensive to use a borosilicate joint and a graded seal.

3. **Glass-to-Metal Seals.** These seals are extremely useful for connecting lecture bottles, thermocouple gauges, copper tubing, etc., to glass manifolds or to glass joints. Conversely, they can also be used for attaching glass parts to metal lines or equipment. Direct seals of borosilicate glass to copper tubing are available.

A highly versatile seal is the Kovar seal which is a graduation from Kovar, an alloy of iron, nickel, and cobalt to a matching glass. The Kovar can be brazed or soft soldered to many other metals. These seals cost about half as much as the direct copper seals and are available from practically every supplier of scientific glassware.

For simple applications such as a simple metal electrode or a flame test wire, metal wires can be sealed into glass tubes or rods. Platinum wire and Nichrome wire are used frequently.

Metal-to-Glass Joints

Almost any metal can be made to stick into a piece of glass. However, if the seal must be vacuum tight or have great mechanical strength, the coefficient of expansion of the metal must match that of the glass within about 1 part per million. Platinum, which expands at 90 parts per 10 million parts, is an excellent match for a soft glass such as Kimble's lime glass. Tungsten can be sealed to

Corning No. 7720 borosilicate glass or Corning No. 3320 uranium glass.

To seal either soft or borosilicate glasses to copper wire, first flatten the wire section that will come in contact with the glass to a thickness of not more than 0.5 millimeter, and then file the edges to knife-sharpness. Heat the metal piece until the color changes to a reddish-brown, indicating the formation of a light oxide film, and immediately paint it with a concentrated solution of borax or drop it into a solution of household detergent containing borax. When dry, the wire should be uniformly coated with a white film of borax. The metal may then be incorporated into a **press seal.** The wire is inserted so the flattened section is just inside the glass tube. Limit the heated glass area to the section that contains the flattened portion of the wire, and concentrate the flame more on the glass than on the metal (Figure 72). Heat converts the borax into a form of glass that not only helps to dissolve the

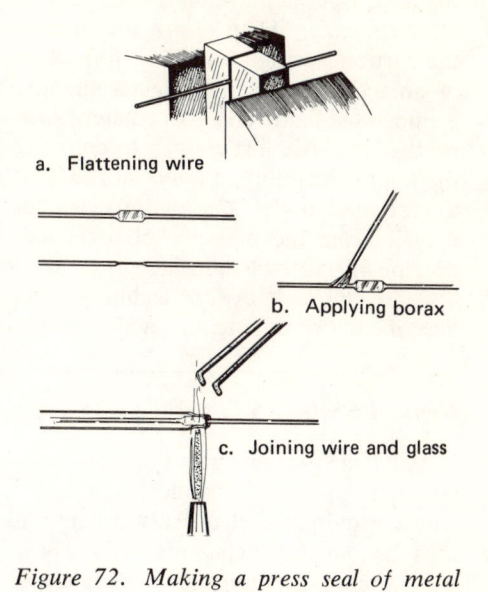

a. Flattening wire

b. Applying borax

c. Joining wire and glass

Figure 72. Making a press seal of metal and glass

metal oxide but shields the copper from excessive additional oxidation. Carefully anneal the seal.

Vacuum Connections and Seals

One useful vacuum connection is a **metal bellows,** a one-piece tube having deeply folded or corrugated sidewalls that permit movement. These connections are indispensible when connecting heavy or vibrating equipment to a glass vacuum system. The bellows acts as a shock absorber and absorbs the vibration before it reaches the glass system. In making this connection to the vacuum system, a Kovar metal-to-glass seal is usually used.

It is often necessary to introduce material into a vacuum system or to remove it without changing the pressure or exposing the material to the air. Both operations can be performed through the used of **break seals** and **seal-offskis.**

A break seal is a tube with an easily breakable partition, such as a drawn point or a thin bubble. A nail or metal weight is enclosed in the upper section of the break seal which is then attached to the vacuum manifold and evacuated. The seal is broken by raising the enclosed weight with an external magnet and allowing it to drop. A seal-offski is simply a thick walled construction which permits sealing under vacuum. Removal of condensation or sublimable material from the vacuum system involves condensing it into a closed tube equipped with a seal-offski and flame sealing at the constriction with an air-oxygen hand torch.

Break seals are available from most of the laboratory supply houses and manufacturers of chemical glassware. Seal-offskis are not commercially available and must therefore be made by the user.

Glass Joints and Stopcocks

Standardized joints for connecting glass apparatus are commonly used in all types of chemical laboratories. When compared with using cork and rubber stoppers, these joints drastically reduce time required for assembling or disassembling apparatus, eliminate contamination, and permit a high vacuum to be maintained. The most common joint is a **Standard Taper** ($\mathbb{T}$) ground joint. The Standard Taper specifications are prepared by the National Bureau of Standards to give uniformity and interchangeability to laboratory glassware. The joint surfaces are ground to give rough looking surfaces that fit together very well. Joints are interchangeable and are designated by two code numbers. The first number gives the diameter of the small end of the ground surface to the nearest millimeter and the second number gives the length of the ground surface at the small end. (*See* Table XIII). Fittings are designated as male and female.

A second type of ground glass joint is the **ball and socket** joint. The two parts are known as the ball and the socket or cup. They are held together by a special

Table XIII. Standard Taper Joints

Inner Part Only	
Designation (Large Diam/Length)	*Straight Tubing End Approx. o.d., mm*
5/20	5
19/38	17
24/40	22
45/50	42
103/60	100
Outer Part Only	
5/20	8
19/38	22
24/40	28
45/50	50

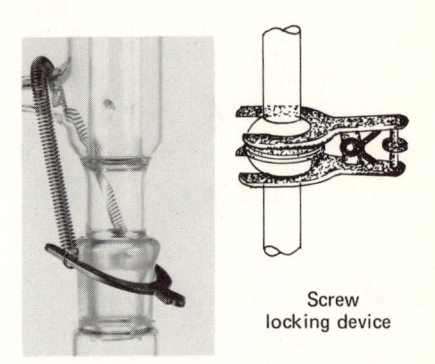

Screw locking device

Figure 73. Methods of securing joints

Table XIV. Ball and Socket Joints

Designation Diam of Ball/i.d., mm/mm	
Ball Only	*Socket Only*
7/1	7/1
18/7	18/7
18/9	18/9
28/12	28/12
50/30	50/30
102/75	102/75

spring-loaded clip (Figure 73). The designation code gives the diameter of the ball first and the inside diameter of the tube second (Table XIV). Systems with ball and socket joints require less precise alignment than ground joint glassware. When a flask must be frequently removed from a system, the ball and socket joint may facilitate the removal or connection.

Other types of joints use **O-rings.** Some glass joints must be precision tooled for an O-ring seal. These give a greaseless, vacuum tight connector which can be quickly assembled and dismantled by using a regular spherical joint clamp. One variation on this form of joint uses a Teflon seal with the O-ring. The joint will not freeze and provides a chemically inert Teflon and glass surface to liquids or gases. (*See* Figure 74).

Ground joint stopcocks are used throughout glass systems to stop or con-

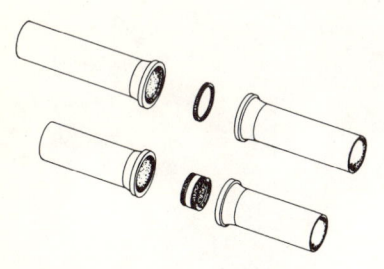

Figure 74. Joints using O-rings

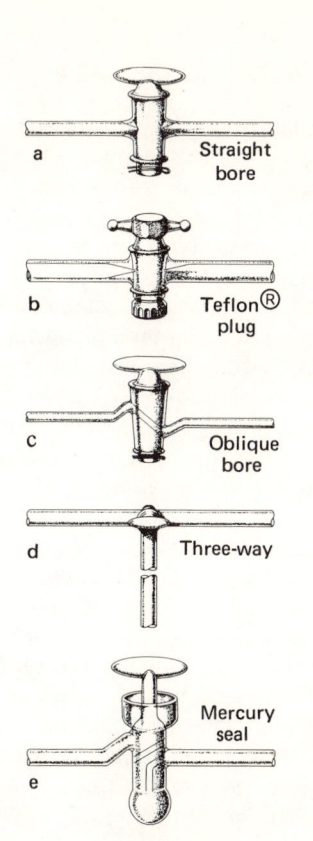

trol the flow of a gas or liquid through the system. The stopcock consists of a glass plug ground to fit into the barrel or body of the stopcock. A hole or several holes are bored through the plug to match precisely the openings in the barrel. Common types of stopcocks are shown in Figure 75. There are a number of different types of stopcocks, each designed for a specific purpose. Vacuum stopcocks, for example, have an oblique bore and the plug is held in place by the vacuum in some designs. **Stopcock parts are not interchangeable unless so designated by a symbol.**

The efficiency of a stopcock under vacuum is not only determined by its design and construction, but also by the type of grease used and by the way it is applied. Perhaps the best method of greasing a stopcock is to apply the lubricant in parallel, length-wise strips on the ground glass plug. After pressing the plug into its seat, the grease spreads and forms capillary paths through which the last traces of air can escape, thereby forming a seal without air bubbles or air streaks. The types of lubricant used will depend on the temperature requirement and on the type of material being handled in the system.

All ground joints must be lubricated (greased) before they can be used, but not all fittings (*i.e.*, Vitrokit) require lubrication. Before lubrication, the surfaces on both male and female parts of

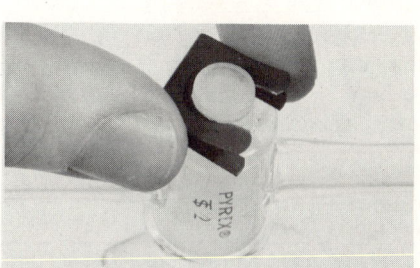

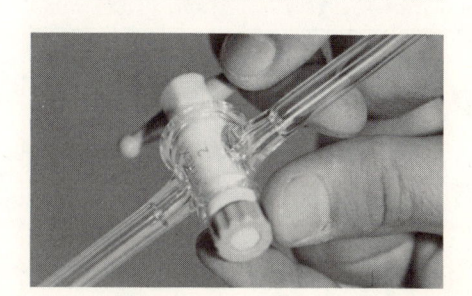

Figure 75. Common stopcocks

the glassware must be perfectly clean and dry. The lubrication in the joint prevents leakage and wear of the fittings. A number of special greases are available for lubricating ground glass joints. (*See* Table XV.) In using lubricants, apply only a light coat over the upper two-thirds of the male joint.

Excess lubricant may block the tube openings, cause leaks in a stopcock, or drip into a reaction vessel and cause unnecessary contamination. Insufficient lubrication may cause fittings to leak or a stopcock to stick. Some practice is necessary to determine the exact amount to use. Every time a ground joint setup is dismantled, the grease should be carefully cleaned from the ground surfaces. Lubricants are usually supplied in tubes.

Greaseless joint apparatus make it possible to construct a completely greaseless vacuum system. Teflon sleeves or Teflon tape can also be used. These fit over the male ground glass joint to eliminate joint freezing and contamination (except from metallic sodium). Moreover, the sleeves are heat resistant up to 300°C. Care should be taken with

Table XV. Lubricating Greases

Catalog Designation	Brand Designation	Use	Temperature Conditions
Stopcock lubricant	Lubriseal, improved formula	For sealing and lubricating stopcock plugs; resistant to acids and alkalies	Melting point, about 40°C
Ground joint grease, vacuum	Apiezon H	A multi-grade grease for use over a relatively wide range of temperatures	Will withstand temperatures up to 250°C; intended for applications from ambient down to approximately —15°C
Stopcock grease, high vacuum	Apiezon N	For lubricating and sealing glass stopcocks used in high vacuum apparatus	Safe maximum temperature in use, 30°C
Soft wax sealing compound	Apiezon Q	For semi-permanent joints; has special utility for the sealing of underground joints	Remains firm at temperatures up to 30°C; safe maximum temperature in use, 30°C
Stopcock grease	Silicone, Dow Corning	An effective sealer and lubricant for glass and ceramic stopcocks; resistant to corrosive chemicals, insoluble in water	For continuous operation at temperatures from —40°C to 200°C
Stopcock grease, high vacuum	Silicone, Dow Corning	Effectively seals and prevents freezing of stopcocks and other ground glass joints in high vacuum systems	Characterized by heat stability

Teflon parts to avoid forcing them together. Teflon has cold-flow properties which tend to distort machined or formed parts. Teflon is the DuPont Company's registered trademark for its fluorocarbon (polytetrafluorethylene) resins, TFE and FEP. Teflon TFE is a high melting thermoplastic material. Teflon TFE, the first resin developed, has the familiar white-to-opaque appearance. These types of resins have a combination of chemical, electrical, mechanical, and thermal properties unmatched by any other material. Because of these properties, Teflon resins are used as gasket and packing materials in chemical processing equipment, as electrical insulation for maximum reliability, and in seals or other mechanical applications including those requiring antistick characteristics. Teflon TFE comes as sheets, rods, tube, tape, bars, and cylinders. The only chemicals and solvents that affect Teflon are molten alkali metals and certain halogenated chemicals at high temperatures and pressures.

Proper alignment is more critical with taper joints than with ball and socket joints. Use as few clamps as necesary for full support and safety. When possible, take advantage of the rigidity of the taper joints for support rather than adding more clamps. Support receiver flasks and similar components from the bottom on rings and wire gauzes. Then assemble from the bottom up, adding clamps only as necessary and finally tightening the clamps only after all components are properly aligned. Depending upon the diameter of the glassware at the point of support, use regular utility C clamps, three-finger, or universal clamps. If a component such as a receiver flask must be replaced from time to time during a reaction or distillation, arrange the system initially so that this can be done easily.

1. Do not scratch mating surfaces with abrasives or the metal wires in cleaning brushes.

2. Keep stopcock plugs lubricated and the orifice free of lubricant.

3. If a joint, stopper, or stopcock freezes and cannot be separated without undue force, ask your instructor or supervisor for help. The instruction manual usually gives directions for loosening frozen joints.

4. Do not leave ground glass components tightly joined any longer than necessary in the presence of alkalies or alkaline solutions. They will almost surely freeze.

5. Inspect all glassware carefully for cracks or dangerous chipping. Do not use a piece of damaged glassware.

6. Round bottom flasks should not be placed on a hard, flat surface; they will probably break if this is done.

7. Insert thermometers very cautiously through the rubber adapters, using water or glycerol to assist insertion. Protect your hands with cloth or gloves.

8. In assembling and clamping the glassware, make the clamps fit the apparatus. Tightening the clamps before all components are properly aligned will cause breakage.

Ordering Supplies

Most laboratory personnel must sometime order supplies, reagents, and apparatus. Some materials may come from an in-house stockroom which has its own catalog. Other materials must be ordered directly from a laboratory supply house or a laboratory chemicals firm. Unfortunately, there is not a single uniform system by which chemicals are packaged. Laboratory supplies are categorized differently among the various

TX733-9*	Toluene, Criterioquality, 99.9+ mol%	5 ml	40.00
TX734*	Toluene, Chromatoquality 99+ mol% Actual lot chromatographic curve is supplied with each unit.	500 ml / 3 l	4.00 / 14.00
TX735*	Toluene, Reagent .... $C_6H_5CH_3$ FW 92.14	1 pt / 12 x 1 pt / 8 pt / 4 x 8 pt / 5 gal	1.10 / 12.00 / 5.04 / 17.92 / 16.80

Meets A. C. S. Specifications

Boiling range (including 110.6°C .... 2°C
Color (APHA) .... 10 max.
Residue after Evapn. .... 0.001 %
Subs. Darkened by H_2SO_4 ...... to pass test
Sulfur Compounds (as S) 0.003%
Water 0.03%

P325	Toluene (Pract.) BP 107-109°	1 kg. / 3 kg. / 5 gal.	3.55 / 4.45 / 13.35

*Figure 76. Sample catalog specifications
for toluene*

supply catalogs. It will be useful to understand the conventions which describe the various ways in which materials are packaged.

Let us look at what is necessary when placing an order for some organic chemicals, mineral acids, inorganic compounds, and several items of hardware. One item needed is toluene. Six chemical catalogs show that toluene ranges in price from $2.67 per gal to $30,400 per gal! Grades of purity offered include: "Ultrex," "Reagent ACS," "Reagent," "Criteriolquality," "Chromoquality," and "Purified." The six chemicals catalogs provided thirty different descriptions for toluene! This example shows the number of choices concerning purity and packaging. These factors are reflected in the price you pay and the quantity you will want. It is important to give attention to all details when ordering materials. Fig-

ure 76 shows some sample catalog descriptions for toluene.

Chemicals are packaged with the container label giving the following information:

1. The product name
2. A code number which conforms with the catalog description
3. Identification for hazardous chemicals such as corrosive liquids, flammables, poisons, etc. Precautionary measures to be followed to protect against the hazard; first-aid treatment where advisable
4. The grade or quality of the chemical

Figure 77 shows a typical label.

Quality definitions for chemicals are given below.

Primary Standard. A specially manufactured analytical reagent of exceptional

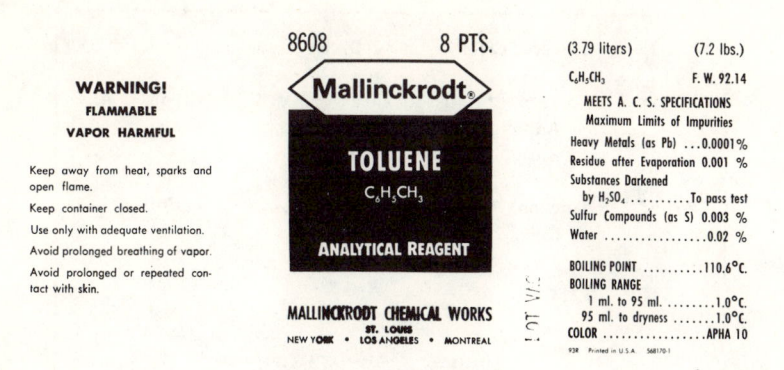

Figure 77. A typical reagent label

purity for standardizing volumetric solutions and preparing reference standards.

Reagent (ACS). Maximum limits of purity for most commonly used reagents (mostly inorganic) have been established by the Committee on Analytical Reagents of the American Chemical Society. (*See* Figure 77.)

Reagent. When the American Chemical Society has not developed specifications for a specific reagent, the manufacturer establishes its own standards and the maximum limits of allowable impurities are shown on the labels of these reagents.

C.P. "Chemically Pure" grades of chemicals are offered by manufacturers. They meet or exceed U. S. P. or N. F. requirements but are of lower grade than "Reagent" or "Reagent ACS" chemicals.

U. S. P. Chemicals labeled U. S. P. meet the requirements of the U. S. Pharmacopoeia. Generally of interest to the pharmaceutical profession, these specifications may not be adequate for reagent use.

N. F. Chemicals labeled N. F. meet the requirements of the National Formulary. Chemicals ordered with an N. F. label may not be useful for reagents. It will be necessary to check the National Formulary in each case.

Practical. This grade designates chemicals of sufficiently high quality to be suitable for use in some syntheses.

Organic chemicals of practical grade may contain small amounts of intermediates, isomers, or homologs.

Purified. A grade of chemical that is physically clean and of good quality but does not meet Reagent ACS, Reagent, U. S. P., N. F., or C. P. standards.

Technical. A grade of chemical generally suitable for industrial use. Purity is not specified and is generally determined by on-site analysis.

Spectro Grade. A designation for organic solvents which have been prepared for use in ultraviolet or infrared spectrophotometry without further purification. Many of these also conform to Reagent or Reagent ACS standards.

Use the following units and abbreviations for ordering chemicals.

Solid chemicals: Pound (lb), Ounce (oz), Kilogram (kg), Gram (g)

Liquid chemicals: Pound (lb), Ounce (oz), Kilogram (kg), Gram (g), Gallon (gal), Pint (pt), Liter (l.), Milliliter (ml)

Use the following chemical container abbreviations. Unless otherwise specified, chemicals are packaged in bottles with appropriate screw caps and liners.

amp in can	ampule in can
bot	bottle
b. & c.	bottle and can
b.m.t.	bottle in mailing tube

box c.	boxed can
c.	can
can pl.	can, polyethylene liner
cby.	carboy
ctn.	carton
fb.pl.b.	fibre, polyethylene-lined box
f. dr.	fibre drum
f.pl.dr.	fibre, polyethylene-lined tube
m.t.	mailing tube
p.	packet
pa.env.	paper envelope
p.d.	plastic dispensers
pkg	package
pl.b. or poly bot	polyethylene bottle
pl.cby.	polyethylene carboy
s.p.	steel pail
scb	screw cap bottle
sb	sling boy

Let us consider another example. A laboratory supply house catalog lists

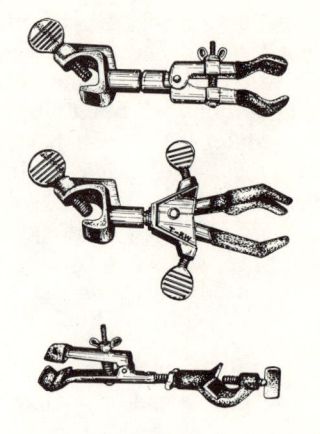

Figure 78. Examples of buret clamps

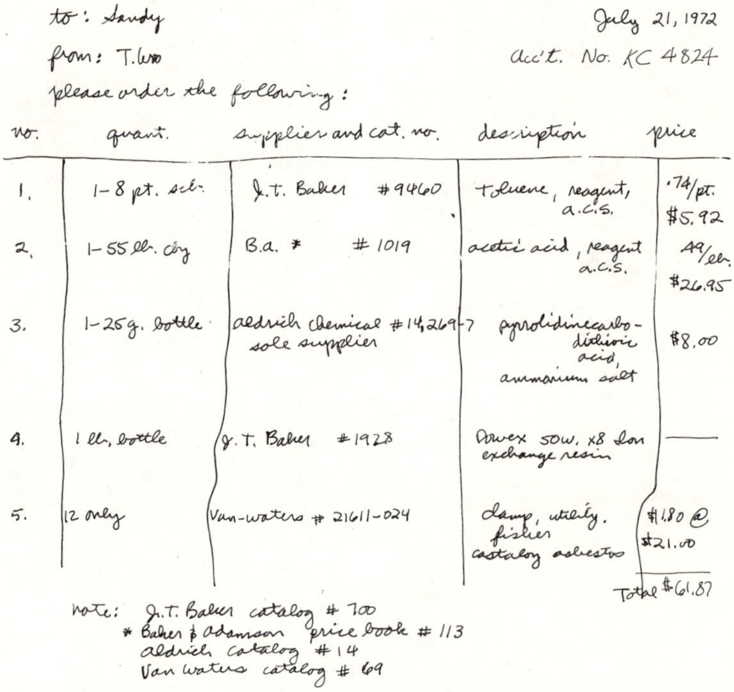

nine different styles for buret clamps. The clamp jaws may be covered with asbestos, vinyl, or rubber. The clamps may be made of Castaloy (a corrosion resistant alloy and a trademark of Fisher

to: Sandy   July 21, 1972

from: T. Wo   acc't. No. KC 4824

please order the following:

no.	quant.	supplier and cat. no.	description	price
1.	1-8 pt. scb.	J.T. Baker #9460	toluene, reagent, a.c.s.	.74/pt. $5.92
2.	1-55 lb. cby	B.a. * #1019	acetic acid, reagent a.c.s.	49/lb. $26.95
3.	1-25 g. bottle	Aldrich Chemical #14,269-7 sole supplier	pyrrolidinecarbodithioic acid, ammonium salt	$8.00
4.	1 lb. bottle	J.T. Baker #1928	Dowex 50w, x8 ion exchange resin	
5.	12 only	Van-Waters # 21611-024	clamp, utility, fisher castaloy asbestos	$1.80 @ $21.00
				Total $61.87

note: J.T. Baker catalog #700
        * Baker & Adamson price book #113
        Aldrich catalog #14
        Van Waters catalog #69

Figure 79. A sample order request

Scientific), stainless steel, or cadmium plated stamped steel. The prices range from $.40 to $1.65. Figure 78 shows the various configurations of buret clamps listed in one catalog. You will have to decide which style and grade to buy, based on the experimental needs.

The secretary, stockroom attendant, or supervisor who will work from your order list will need to know the following:

1. The quantity of each item needed.
2. The units in which the item is packaged.

3. The manufacturers' order number.
4. The manufacturers' catalog description.
5. The cost per unit ordered.
6. The total cost of the order.
7. The catalog number from which the description was obtained.

A sample order sheet may look like the one reproduced in Figure 79. It is good practice to make copies of all your paper work. A personal desk or file copy of an order is frequently very useful.

Specifications Testing

The ability to describe the physical and chemical properties of materials, particularly to regulate quality for buying and selling, is extremely important in operating all industrial and governmental operations. Chemical technicians have a large and important role in assuring that material bought or sold by a company or agency meets the specifications required by industry or government. A **specification** describes the qualities of a material so that two or more individuals can agree that if the material meets the specification, it will be satisfactory for the intended use.

Specifications

We all buy materials by specification. When we ask the garage attendant to fill the crankcase with 30-weight oil, we are really saying that we want to use a grade of oil which is described by the Society of Automotive Engineers (SAE) as having a viscosity of 30 when tested by a particular piece of equipment—the Saybolt viscometer. The value of that specification is that you can go to any service station of any oil company and buy any 30-weight oil with assurance that the oil does meet the description of SAE 30. The oils may differ in ways not covered by the specification. For example, better oils may contain corrosion inhibitors

or antioxidents, but all of them will meet the above specification.

If a steak in a supermarket is stamped U.S. Choice, we know that the meat has passed a U.S. Department of Agriculture (USDA) inspection and meets the standard for fat content and distribution, texture, color, and water content required in the specification for the U.S. Choice grade.

The SAE and USDA represent two types of organizations that produce material specifications. SAE is an organization supported by many industrial corporations interested in testing, specifications, and uniformity of automotive products. USDA is interested in protecting consumers through the inspection, testing, and specifications of agricultural products which are shipped across state lines. Other sources of specifications are individual companies, states, or cities which produce specifications for their own use.

As a typical example, the Eastman Kodak Co. of Rochester, New York, uses a yellow-orange color to identify its products. This color appears on all of its boxes, labels, advertising, film cassettes, books, and brochures. The color is identical whether the material was made in Rochester, or printed in a plant in Chicago, or whether it appears on a cassette made in Germany, England, or Japan. The Kodak company obviously has a very strict and accurate specification for this color which must be followed by all of its plants, suppliers, and advertising printers.

Specification Procedures

Much of the analytical work done by technicians in laboratories is specifications testing, using official methods. The work is very important since large amounts of money and materials are exchanged on the basis of products meeting specifications.

One particular characteristic which must guide all of your actions and judgements while performing specifications testing is that you must follow the procedure *exactly* as it is stated in the specification—even though your scientific judgement may suggest a better, or even more accurate, way to do the job. What does a specification look like? Generally, it is a printed form in an individual brochure or book, and would typically look like the one in Figure 1.

Let us examine this particular specification for Effectiveness of Corrosion Inhibitors.

First we see that it is a UOP method. This means that it is recommended by Universal Oil Products Corporation, a large petroleum process design company. The use of the test is given by the scope, which also says that the test is similar to that recommended by two other very important specifications sources—the America Society for Testing and Materials (ASTM) and the U.S. Department of Defense (MIL).

The specification then outlines the purpose of the test. It states, either by reference to an ASTM specification or giving details here, the exact apparatus and reagents required. If we were doing the test, we would try to do it with exactly those materials described.

The specification gives an exact, step-by-step procedure, then the method of

calculating and reporting the result. An illustration of the test apparatus is shown also.

Specification Sources

Over the years, many organizations and agencies have, individually or in groups, created specifications and tests which are generally recognized and accepted. If you are employed by a large company or agency laboratory, the laboratory and library would undoubtedly have on file the specifications for products with which the company commonly deals.

Some of the organizations which issue specifications are described below.

ASTM. American Society for Testing and Materials. This organization is supported by a vast number of industrial organizations. Its membership consists of scientists, technologists, and engineers who serve on committees which try to create better methods for testing and specifying many classes of industrial materials.

Products which are generally covered by ASTM specifications include: metals, cement, concrete, road surfacing, masonry, ceramics, insulation (thermal, acoustic, and electrical), wood, paper, adhesives, leather, petroleum products, naval stores, coal, coke, gaseous fuel, antifreeze, plastics, rubber, textiles, soap, wax, water, and air.

An important, additional function of ASTM is to provide standard operating methods and specifications for instrumentation. ASTM committees publish calibration procedures for emission and absorption spectrometers, gas chromatographs, thermal analyzers, and mass spectrometers. These procedures state the test conditions and results that should be obtained in order to assure that your instrument is working properly. They define the materials to be used as stand-

Effectiveness of Corrosion Inhibitors by a Modified Turbine Oil Rusting Test

UOP Method 354-58

SCOPE

This test is used to determine the minimum effective concentration of corrosion inhibitors in military aircraft fuels. It is a modification of the ASTM Method D 665-54[1] and MIL-I-25017 (ASG)[2] using corrosion probes described by Marsh and Schaschl.[3] It may be extended to other hydrocarbon fuels.

OUTLINE OF METHOD

This method determines the corrosion of a 1-mil strip of steel after 20 hours of immersion in inhibited and uninhibited mixtures of isooctane and synthetic sea water. The change in the electrical resistance of the strip is taken as the measure of corrosion. A temperature-compensating probe of special design is employed to carry the corrosion strip.

APPARATUS

The apparatus used is as described in ASTM D 665 except that the large hole in the beaker cover is enlarged to 1 inch diameter on the same center. The hole is to accommodate a No. 5 neoprene stopper.

Test specimen. A special type temperature-compensated corrosion probe made from 1-mil cold rolled 1020 steel shim stock. See drawing for details. The probe consists of a steel strip with insulated copper leads attached at both ends and at the middle. This forms 2 legs, or half, of a resistance bridge. Half of the strip, 1 leg of the bridge, is enclosed in a glass tube to protect it from the sample. This leg provides temperature compensation. The other leg is exposed to the sample and measures corrosion (see Note 1).

Corrosion meter, Labline-Pure, Corro-Dex, Model 5205, or equivalent.[4]

REAGENTS

Acetone, CP
Etching acid, a 1:1 mixture of concentrated hydrochloric acid and distilled water containing 1% ferric chloride in the mixture.
Isooctane, knock test grade, depolarized by passing through a silica gel column.
Synthetic sea water, as described in ASTM D 665
Toluene, CP

PROCEDURE

The inhibitor is added to 350-ml portions of isooctane in clean separatory funnels such that concentrations of 0, 10, 20, 30, 40 and 50 ppm of inhibitor are obtained. The stopcocks of the separatory funnels must not be greased or lubricated. Add 35 ml of distilled water to each funnel. Shake each funnel 1 minute and allow to settle. Drain off the water layer.

Place 300 ml of the washed isooctane solution in a beaker and set the beaker in the 100 F bath. Put the stirrer and beaker cover in place. When all the beakers are filled, start the stirrers.

Prepare the probes while the beakers are warming. Rinse a probe in toluene and then in acetone. Fit the probe into a suitable bottle which is almost full of the etching acid. Shake the probe in the acid 45 times at a rate of 1 shake per second. This exposure to the acid should remove approximately 5 microinches of steel from the surface of the probe. This may be verified from time to time with the corrosion meter. Immediately rinse the probe well with distilled water, acetone and toluene, in that order. Avoid exposure of the metal to air as much as possible (see Note 2). Immediately place the freshly prepared probe into a test beaker before proceeding to prepare the next probe.

Put the probes through the beaker covers, setting them so that the broad side of the exposed metal ribbon faces into the direction of stirring. The probe wires may be bent forward for convenience in connecting the corrosion meter. Let the probes soak 30 minutes in the stirred isooctane, during which time the initial resistance reading of each probe is taken and recorded.

After the 30-minute soaking period add 30 ml of synthetic sea water to the bottom of each beaker while stirring continues. Read and record the resistance of each probe at about 4-hour intervals and after exactly 20 hours. If desired, the probe may be washed with distilled water and acetone. The exposed metal strip then may be cut out and preserved in the data book under transparent tape.

CALCULATION AND REPORT

Compute the microinches of corrosion for each probe at 20 hours (see Note 3). Plot the microinches of corrosion at 20 hours against parts per million inhibitor in the isooctane. The minimum effective concentration may be considered as that concentration of inhibitor at which the corrosion consistently averages less than 5 microinches at 20 hours.

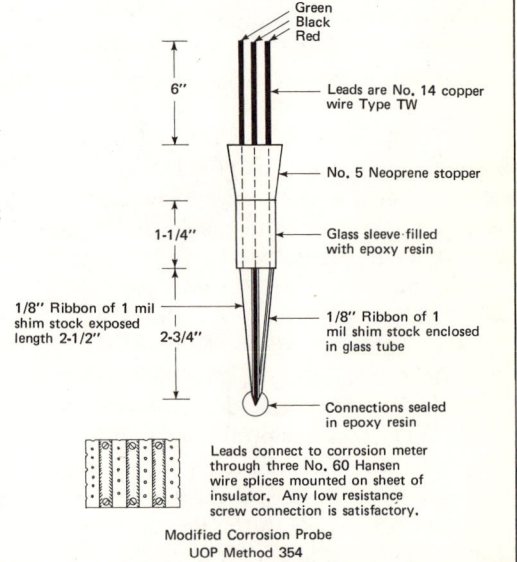

Modified Corrosion Probe
UOP Method 354

Figure 1. Example of a specification

ard test samples and precisely state the definitions and terms used in each instrumental technique.

AOAC. The Association of Official Analytical Chemists. This organization publishes volumes of accepted analytical methods and a journal which describes new analytical methods being considered for acceptance. Most AOAC specifications apply to agriculture-related materials while ASTM methods are mainly applied to industrial products. AOAC tests include food, drugs, tobacco, pesticides, and a large number of microbiological methods as well as chemical and physical tests.

USDOD. U.S. Department of Defense. The specifications for millions of items purchased by the military are defined by the Defense Department. One typical agency that produces many specifications of importance to chemical technicians is the U.S. Army Natick Laboratory. A typical specification states the requirements for fabric to be used in making flags. The specification lists requirements and test methods for color, chemical identity, melting point, denier, weave, weight per square yard, yarns per inch, tensile strength, tear strength, air permeability, and shrinkage. USDOD specifications are usually shown in literature with the prefix MIL (for military). The specifications for flag fabric is MIL–606D.

FDA. U.S. Food and Drug Administration. The FDA, unlike the Department of Defense, does not issue specifications to govern purchasing of materials by the government, but rather to regulate the quality and safety of products produced and sold commercially in the United States. The objective is to protect the consumer from unsafe, misrepresented, or useless food, drug, and cosmetic products. Other agencies producing specifications of interest to chemical technicians for regulatory purposes include the Department of Agriculture, the Interstate Commerce Commission, and the National Bureau of Standards.

USP. United States Pharmacopoeial Convention. This organization publishes the *U.S. Pharmacopoeia,* a volume containing descriptions and test procedures for hundreds of ethical and proprietary drugs made and sold by the pharmaceutical industry in the United States. You will often see chemicals in the laboratory with a label such as Boric Acid—USP.

This indicates that the material meets the specifications stated in the *U.S. Pharmacopoeia.*

NF. National Formulary of the American Pharmaceutical Association. These specifications cover a range of drug products similar to those in the USP. Drugs conforming to these specifications are labeled NF.

ACS. The American Chemical Society. This is a very important source of specifications for fine chemical quality. ACS standards will often be shown on the labels of reagent grade chemicals used in the laboratory. If you are employed in the laboratories of a reagent manufacturer, these specifications would be important in your work.

Should you work in a laboratory concerned with the quality of some important industrial or agricultural product, trade organization specifications beyond those of the large organizations above will probably be used. Special specifications cover the quality of such things as wool, diamonds, fur, and oil.

For example, the set of specifications issued by the Wool Textile Association of London, defines very extensively the color, size, weight, curl, tightness, and fiber length, as well as chemical properties of wool. Since wool is often an item of international trade, these specifications are used wherever wool is produced to assure a uniform measure of quality.

In addition to these national and international sources of specifications, states and cities also issue specifications covering products made and used within the state or purchased for use by agencies of the government. For example, California issues specifications for the testing and requirements for air pollution control of automobile exhausts. These specifications apply only to California vehicles and often are different from those required in other states.

Chemical Literature and the Library

As a chemical technician, you must read and understand instructions, keep a useful notebook, produce reports, and use a library advantageously. Familiarity with chemical library and the literature it contains is just as important as any other general skill of a superior chemical technician. Time spent in the library can reduce time spent in the laboratory by many hours. Frequently, it produces a satisfactory solution to a problem with no time wasted in repeating what someone else has already done and described.

From this chapter you will learn to use library facilities intelligently and rapidly. You should not need to ask the librarian repeatedly to search for you. The librarian is there to help and to answer questions when you get stuck. The references listed at the end of this chapter will provide help for more extensive searches.

Before getting acquainted with the detailed location of books in the library, you should be familiar with the kinds of material available and how it is listed. You will then be less likely to waste time by looking in the wrong places.

Categories of Published Material

Scientific literature is broadly divided into two main groups—journals and reference works. Journals are also divided into two groups—primary and secondary. Except for reference works, these groups are not mutually exclusive. Journals may be either primary or secondary or both. While books are usually secondary, some, particularly proceedings and symposium books, contain primary material. Reference works, however, are usually books.

Primary Sources. The term, "primary," doesn't mean "best," nor does it mean that a primary journal is the first place to look for an answer. A primary journal is simply a periodical that publishes first reports of discoveries—of new compounds, processes, or theories—that are the results of scientific research. Such journals have two general types of material: formal papers that report research results in detail and communications that are flash announcements of new discoveries. Communications serve mainly to establish priority of discovery and may be important in patent work. Formal papers usually follow with details.

Publication of research papers is not automatic. A paper has to pass inspection, first by the editor of the journal, then by reviewers familiar with the subject who are selected by the editor. Publication, therefore, means that the paper has been approved, and the mere fact that it appears in print stamps it as being important and authoritative (although not always infallible). Some journal editors are very strict and reject as many as half the papers submitted to them. Others are more lenient, and some pass papers that contain speculative material. You have to get to know the journal to know how to judge its material. Because of their very nature—as vehicles for reports on new research and ideas, primary journals are not highly organized. They usually contain a mix of papers within a given subject area that are published in the order in which they are received. Thus, they are not the places to start looking for specific information. Unless you have been assigned special projects of advanced research, you will probably consult original papers only after being referred to them by a secondary source.

Secondary Sources. Secondary literature includes all the other kinds of published information such as secondary journals and most books—textbooks, reference books, handbooks, and most books of proceedings. To say that a work is "secondary" means simply that it contains information from the primary sources—collected, organized, and usually evaluated. It contains no new data. Secondary sources digest the great masses of material from the primary sources into a more usable form and make it easier to find information. The disadvantage is the time lag between publication of primary and secondary material. However, you will probably find secondary sources, with their organized presentations, much more useful than primary sources. You should never plow through primary sources to find data that is readily available in a handbook!

Kinds of Publications

Following are the major types of published material, both primary and secondary, that will be useful in a literature search. Fine distinctions cannot always be made because some publications include several types. Perhaps you will not use each of the categories of published material in the order listed, but you should recognize the style and purposes of each one.

Original Papers and Articles. These publications cover new material (research results, etc.) not previously described. They are the core of communicating new advances, and their importance cannot be overemphasized. These papers or articles are usually published in periodicals or journals which cover a limited area, such as *Journal of the American Chemical Society, Journal of the Chemical Society* (Great Britain), *The Journal of Organic Chemistry, Analytical Chemistry, Journal of Chromatographic Sciences,* and hundreds of others in many languages. Some journals such as *Science, Nature,* and the *Journal of Chemical Education* frequently include other types of papers.

Patents. Patents have already been mentioned in Chapter 10. They establish priority of disclosure and can be primary sources of information. By the time a patent is issued, the patentee or the company to which it is assigned has advanced applications well beyond the disclosures in the patent. Patents are important to any industry, but most likely you will rarely have occasion to consult them. There is a governmental publication service which lists newly issued patents. Copies of patents can be obtained at reasonable cost.

Abstracts. Journals of abstracts publish summaries of articles, patents, bulletins, and publications as soon after the original publication as possible. Each abstract ordinarily furnishes the following information: title of the work, author's name, original reference (that is, number or name of the bulletin), volume, page, and year of journal; name of patentee, country, date, and number of patent; or other information which would enable you to locate the original; and usually a brief summary of the paper.

Without abstract journals, a literature search would be hopeless. Since the abstracts provide a complete index, there is little likelihood of overlooking an important article. One disadvantage is that the lag between publication of the original article and of the abstract may vary from a few months to a year.

The most important abstract service for chemical technicians is *Chemical Abstracts,* which is published by the American Chemical Society. It is well indexed, has additional, combined indexes published in separate volumes semiannually, and includes cumulative index volumes covering longer periods of time. For maximum usefulness, a more detailed reference should be studied carefully. You should develop skill in using this type of reference by examining the abstracts and indexes. More information about using *Chemical Abstracts* appears later in this chapter.

A quick way to find articles before they are abstracted in *Chemical Abstracts* is to check the weekly *Chemical Titles,* published by Chemical Abstracts Service. This is a KWIK index (key words in context) of titles of research articles that have been received by *CA* for abstracting. Code numbers at the ends of lines refer to the table of contents in the back of each issue. *CT* has been published since 1962 and is also useful for retroactive searches.

Data Compilations. These are organized presentations of data, usually in tabular form. One of the largest sources of compiled data is the National Standard Reference Data System which is administered by the Office of Standard Reference Data in the National Bureau of Standards, Washington, D.C. It provides optimum access to critically evaluated, compiled data in seven areas of the physical sciences. There are four information services available, including a direct inquiry service.

A comprehensive collection of data is the "International Critical Tables," published by McGraw-Hill. Many other data compilations are in books, but some secondary journals carry data collections. *Journal of Chemical & Engineering Data* is a primary publication devoted entirely to data, and many of its articles report the work of single authors. *Journal of Physical and Chemical Reference Data* presents compilations of critically evaluated data on physical and chemical properties.

Handbooks. In these books, data is sifted, and only selected values are given. They are the first places to look when you want data. Among the most popular is the "Handbook of Chemistry and Physics" that is published annually by the Chemical Rubber Co., Cleveland. Reasonably recent editions belong in your personal library as they contain enormous amounts of data as well as mathematical tables. The CRC has also published a supplement to their handbook, "Tables for Identification of Organic Compounds."

Two other popular and useful handbooks are the Lange "Handbook of Chemistry" and the Perry, Chilton, Kirkpatrick "Chemical Engineer's Handbook," both published by McGraw-Hill, New York.

An extremely valuable handbook is "The Merck Index" now in its 8th edition (1968), published by Merck & Co., Rahway, N.J. It is particularly helpful for giving characteristics of chemicals and drugs, sorting out tradenames, and giving information on structure.

The "SOCMA Handbook: Commercial Organic Chemical Names," published by Chemical Abstracts Service, presents the structure of and data on pure organic compounds, polymers, and mixtures.

Reviews. These are the basic fare of the secondary literature and usually are the first places where material in the primary journals is summarized and digested. They appear in periodicals of all kinds—review journals, trade magazines, newsmagazines, and even in primary journals. They also appear in books and in serial publications, such as *Annual Reviews of ------, Advances in ------, Progress in ------, Proceedings of ------*, etc. Symposia usually include some review papers as well as reports of original research and may be published in journals (either primary or secondary), serials, or as books.

Bibliographies. Lists of references to the subject of your search may be available. Such lists are included in review articles, which may be found through *Chemical Abstracts*. Don't forget the files of research reports in your company office; related work done before may have information you can use. "The Literature of Chemical Technology," ADVANCES IN CHEMISTRY SERIES, No. 78, discusses the literature of many industrial fields and provides extensive bibliographies in each area.

Monographs. These publications summarize and evaluate all published information on a selected and often limited area. They present the background and development of a topic and the authors' interpretations of the state of the art, including theories and procedures as well as data. A monograph should represent the finest and most authoritative summary of refined knowledge on its topic. If you can find a recent one in the area of your search, you will be fortunate.

Treatises. These are larger collections that usually amount to groups of monographs and are often multivolumed. An example is "Treatise on Analytical Chemistry" by Kolthoff and Elving that is organized in three parts—techniques, elements, and sample type—each with many volumes. "Technical Methods of Analysis" by R. C. Griffin, published by McGraw-Hill, is extremely useful to the practical chemist. Two others that are helpful in the laboratory are "Organic Syntheses" and "Inorganic Syntheses." They are published by John Wiley & Co. as serials with occasional collected volumes. "Reagent Chemicals," 5th edition, 1974, published by the American Chemical Society, compiles ACS specifications for compounds and/or elements and includes new analytical techniques.

Encyclopedias. These are collections of short articles that cover entire disciplines. A good, general purpose encyclopedia can be very helpful for many searches. More specialized works are Van Nostrand Reinhold's "International Encyclopedia of Chemical Science" (1964), "Kingzett's Chemical Encyclopedia" (1966), and "The Encyclopedia of Chemistry" edited by Clark and Hawley.

Encyclopedias are available for many special subdivisions of chemistry such as electrochemistry, biological sciences, spectroscopy, microscopy, plastics, x-rays and gamma rays, process equipment, and analysis. "Encyclopedia of Chemical Technology" by Kirk and Othmer (Wiley-Interscience) is in its second edition and has 22 volumes.

Dictionaries. A number of chemical scientific, and medical dictionaries are available. "The Condensed Chemical Dictionary," edited by G. Hawley, 8th edition, Van Nostrand Reinhold, 1971, "The Dictionary of Commercial Chemicals" by Snell and Snell, and "Hackh's Chemical Dictionary" are particularly useful. Also, make a habit of using a good dictionary of the English language.

Government Publications. These are a prolific source of information on many and varied subjects. You are already familiar with the names of the Patent Office and the National Bureau of Standards. There are virtually hundreds of bureaus, commissions, and divisions under the departments of Agriculture, Defense, Interior, and Health, Education, and Welfare, and other government departments as well as semi-independent or independent establishments within the government such as National Institutes of Health, National Science Foundation, Energy Research and Development Administration, Environmental Protection Agency, Smithsonian Institution, National Research Council, and many others.

The best way to benefit from these sources is to check the monthly catalog of publications distributed by the Superintendent of Documents, Washington, D.C. 20402. Your company library probably gets it. Some of the most useful guides and indexes, however, must be obtained directly from the department or office, some of which publish research reports and surveys in their own journals and books.

Trade Journals. Many professions, trades, manufacturing specialties, and products have magazines that are devoted to feature articles, news, and product and process information on their specialty. Most are published commercially, but professional societies and trade associations also sponsor them. Many industrial firms publish house organs that carry informative and promotional articles on their products and services. Any of these periodicals in your areas of interest are worth seeking out and keeping up with.

For news of the chemical industry, read *Chemical & Engineering News,* published by the American Chemical Society, and *Chemical Week,* published by McGraw-Hill. For popular, broad coverage of science, read *Scientific American,* which has articles on mathematics, chemistry, physics, biology, medicine, astronomy, psychology, and other sciences.

Catalogs, Bulletins, Advertisements. These publications are frequently the first source of news on new chemicals, equipment, and instruments. They are designed to sell products and should be read with that in mind. However, many technical bulletins could be helpful additions to your personal library. Most laboratories and industrial plants keep files of catalogs and bulletins on products and equipment that they use.

Organization of Library Facilities

When you have identified the sources you need to consult, you must know how to find them in the library. Most books are shelved in order of their call numbers. Call number designations are described later, but books will be arranged either alphabetically, from A to Z, or numerically, from 000 to 999. You must get the call number from the card catalog (*see* below) before the book can be found.

Frequently-used books are often put in a special reserve or reference area. Here you will find encyclopedias, handbooks, dictionaries, treatises, bibliographies, indexes, and other similar reference materials. Usually they must be used in the immediate area, but sometimes they may be signed out overnight. The index card in the card catalog will indicate restrictions on the use of these books.

Periodicals are handled differently. They are commonly stored in separate sections. From time to time they are bound in hard covers, a year's issues at a time, leaving only the most recent issues in the original covers. This practice protects the periodicals but is somewhat expensive so that only more heavily-used periodical collections may be bound by some libraries.

Of growing importance is the storage of material on film or fiche. This storage technique requires special retrieval and reading systems. You may find that your library also has built into its film handling facilities a method of making copies of filmed material. These systems are so varied that the best practice is to become familiar with the system you will use. In order to have a good working acquaintance with the system, you need to know how the material is indexed, retrieved, read, copied, and returned.

Classification Systems and Call Numbers

Nearly all libraries follow one of two different classification systems. This allows non-fictional books or publications to be numbered systematically for purposes of cataloging, shelving, and keeping publications on similar subjects together. These are called the Library of Congress and the Dewey Decimal Classification Systems. Abbreviated outlines of these two systems are shown in Table I. Table II lists more detailed excerpts from the pure and applied science subsections. Expanded descriptions of the systems are available in any library.

From these lists you can interpret call numbers. They describe the subject,

provide for systematic shelving and, with additions such as Ref., C., Per., or Pam., can tell you that the material is not in the regular open stacks, but is to be found in the Reference Room, the Case Shelves, the Periodical File, or the Pamphlet Files. The call number is put on the spine of the book as illustrated in Figure 1.

Figure 1. Placement of call numbers on books

Card Catalogs

The library's card catalog is to the library what an index is to a book. It is made up of the card lists, usually contained in a series of file drawers, which describe the books and materials available in that library. This card catalog may not necessarily be the starting point for a literature search, but it is the best reservoir of information about what you can find in that library.

Each book or reference work is represented by three or more cards, one for title, one or more for subject, and one or more for author(s) and/or editor(s). The cards are distinguished from each other by their format, but most libraries file all these cards in one alphabetical

Table I. Library Classification Systems

Library of Congress Classification System

A	General works	PS	American
B–BJ	Philosophy	Q	Science
BL–BX	Religion	QA	Mathematics
C, D	History (Europe, Asia, and Africa)	QB	Astronomy
		QC	Physics
E, F	History (America)	QD	Chemistry
E	General	QE	Geology
F	Local U. S., Canada, Latin America	QH	Natural history
		QK	Botany
G	Geography—anthropology	QL	Zoology
H	Social Sciences	QM	Human anatomy
HB–HJ	Economics	QP	Physiology
HM–HX	Sociology	QR	Bacteriology
J	Political science	R	Medicine
K	Law	RT	Nursing
L	Education	S	Agriculture
M	Music	T	Technology
N	Fine arts	U, V	Military and naval science
P–PZ	Literature	Z	Bibliography
PR	English		

Dewey Decimal Classification of Books

000	**GENERAL WORKS**	560	Paleontology
010	Bibliography	570	Biological sciences
020	Library science	580	Botany
030	General encyclopedias	590	Zoology
050	General periodicals	600	**APPLIED SCIENCE**
100	**PHILOSOPHY**	610	Medicine
200	**RELIGION**	620	Engineering
300	**SOCIAL SCIENCES**	630	Agriculture
330	Economics	640	Home economics
350	Public administration	650	Business
370	Education	660	Industrial chemistry
400	**PHILOLOGY** (Language)	670	Manufacturers
500	**PURE SCIENCE**	680	Mechanic trades
510	Mathematics	690	Building
520	Astronomy	700	**ARTS: RECREATION**
530	Physics	800	**LITERATURE**
540	Chemistry	900	**HISTORY**
550	Earth sciences		

sequence. Occasionally, however, author cards are filed separately, and some large libraries have many special catalog card files.

Figures 2 and 3 show all three types of cards. Notice that the cards are iden-

tical, except for headings on the title and subject cards.

1. The title card is made from an author card by typing or printing the book title in upper and lower case letters directly over the author's name.

Table II. Chemical Entries in Library Classification Systems

Dewey		Library of Congress
530	Physics:	QC
535	Light	QC 351—495
535.84	Spectroscopy	QC 451—467
536	Heat	QC 251—338
537	Electricity	QC 501—771
540	Chemistry:	QD
540.92	Biography of chemists	
541	Theoretical and physical	QD 453—655
542	Practical and experimental	QD 43—64
542.1	Chemical laboratories	QD 51
543	Analytical, general	QD 71—80
544	Analytical, qualitative	QD 81—100
545	Analytical, quantitative	QD 101—150
546	Inorganic	QD 151—199
547	Organic	QD 241—499
548	Crystallography	QD 901—999
549	Mineralogy	QE 351—399
550	Geology:	QE
570	Biology, general:	QH 301—705
578	Microscopy	QH 201—277
600	Useful arts:	T
608	Patents, inventions	T201—339; TP 210
610	Medicine	R
614	Public health	RA
614.09	U. S. Public Health Service	RA 11
615	Therapeutics and pharmacy	RS
620	Engineering, general:	TA
621.35	Chemical electricity	QC 603—605
628	Sanitary engineering	TD
630	Agriculture:	S
630.24	Agricultural chemistry	S 583—588
640	Domestic economy:	TX
641.1	Food, chemically considered	TX 501—612
660	Chemical technology:	TP
661	Chemicals	TP 200—248
662	Explosives	TP 268—299
663	Beverages	TP 500—618
664	Foods	TP 370—465
665	Oils, gas	TP 343—360
666	Ceramics, glass	TP 785—889
667	Bleaching, dyeing	TP 890—931
668	Other organic industries	?
669	Metallurgy, including assaying	TN 550—799
670	Manufactures, articles made of:	TS
671	Metals	TS 200—770
679	Celluloid; plastics	TP 986
690	Building:	TA; TT
691	Materials, processes	TA 401—492

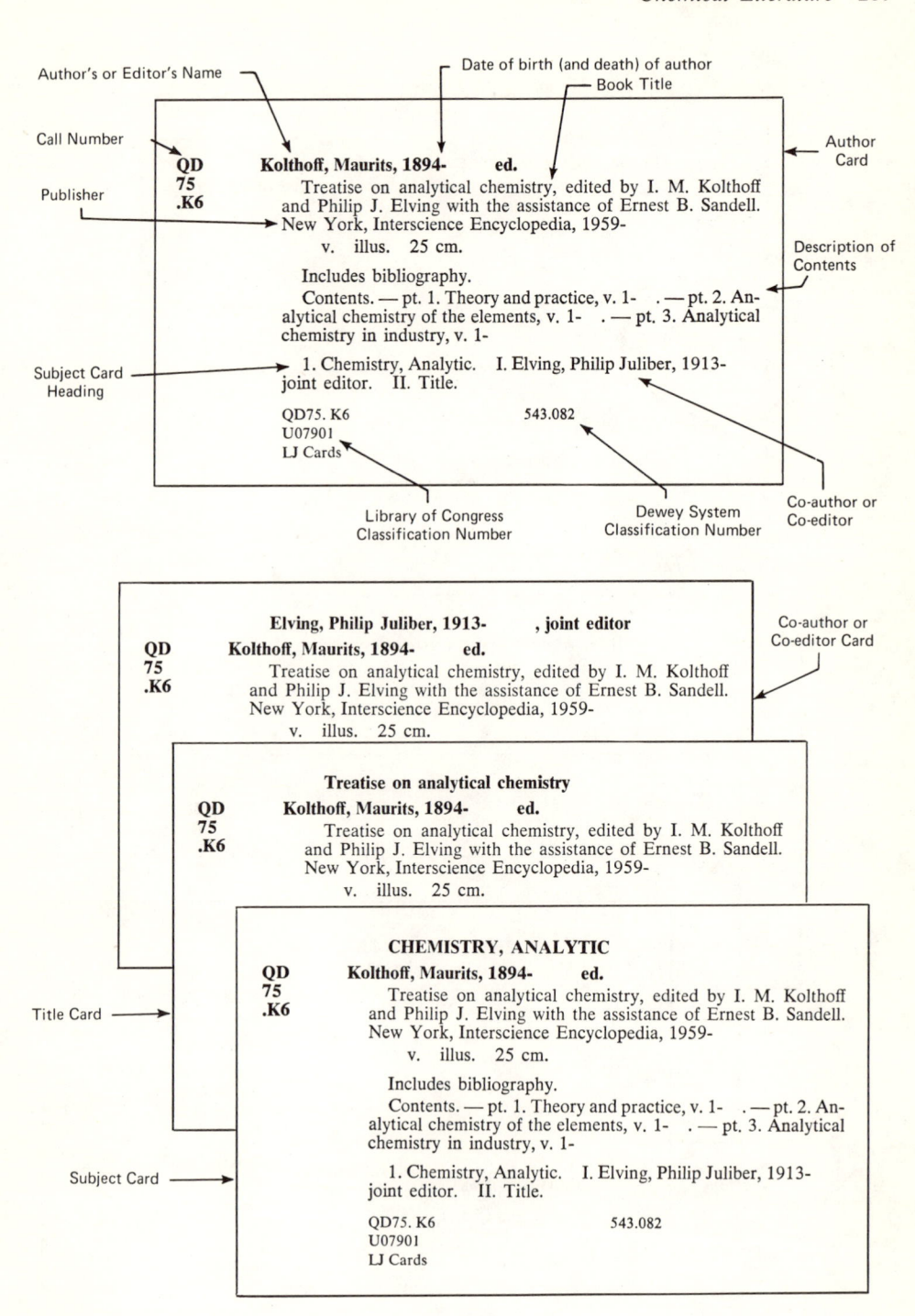

Author's or Editor's Name — Date of birth (and death) of author — Book Title

Call Number

Publisher

QD
75
.K6

Kolthoff, Maurits, 1894-   ed.
   Treatise on analytical chemistry, edited by I. M. Kolthoff
and Philip J. Elving with the assistance of Ernest B. Sandell.
New York, Interscience Encyclopedia, 1959-
      v.   illus.   25 cm.

   Includes bibliography.
      Contents. — pt. 1. Theory and practice, v. 1- . — pt. 2. An-
alytical chemistry of the elements, v. 1- . — pt. 3. Analytical
chemistry in industry, v. 1-

   1. Chemistry, Analytic.   I. Elving, Philip Juliber, 1913-
joint editor.   II. Title.

QD75. K6   543.082
U07901
LJ Cards

Author Card

Description of Contents

Subject Card Heading

Co-author or Co-editor

Library of Congress Classification Number

Dewey System Classification Number

Elving, Philip Juliber, 1913-   , joint editor
QD   **Kolthoff, Maurits, 1894-   ed.**
75   Treatise on analytical chemistry, edited by I. M. Kolthoff
.K6   and Philip J. Elving with the assistance of Ernest B. Sandell.
          New York, Interscience Encyclopedia, 1959-
             v.   illus.   25 cm.

Co-author or Co-editor Card

Treatise on analytical chemistry
QD   **Kolthoff, Maurits, 1894-   ed.**
75   Treatise on analytical chemistry, edited by I. M. Kolthoff
.K6   and Philip J. Elving with the assistance of Ernest B. Sandell.
          New York, Interscience Encyclopedia, 1959-
             v.   illus.   25 cm.

Title Card

Subject Card

CHEMISTRY, ANALYTIC
QD   **Kolthoff, Maurits, 1894-   ed.**
75   Treatise on analytical chemistry, edited by I. M. Kolthoff
.K6   and Philip J. Elving with the assistance of Ernest B. Sandell.
          New York, Interscience Encyclopedia, 1959-
             v.   illus.   25 cm.

   Includes bibliography.
      Contents. — pt. 1. Theory and practice, v. 1-   . — pt. 2. An-
alytical chemistry of the elements, v. 1-   . — pt. 3. Analytical
chemistry in industry, v. 1-

   1. Chemistry, Analytic.   I. Elving, Philip Juliber, 1913-
joint editor.   II. Title.

QD75. K6   543.082
U07901
LJ Cards

Figure 2. Catalog cards, Library of Congress classification system

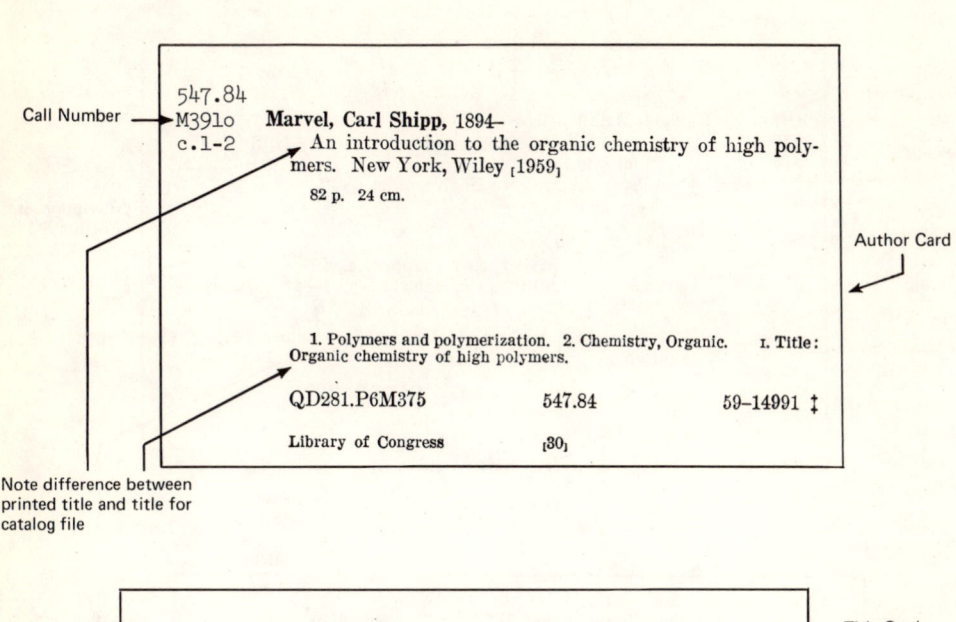

Call Number →

547.84
M391o
c.1-2

Marvel, Carl Shipp, 1894–
    An introduction to the organic chemistry of high poly-
mers. New York, Wiley ₁1959₎

     82 p. 24 cm.

Author Card

1. Polymers and polymerization. 2. Chemistry, Organic. ɪ. Title:
Organic chemistry of high polymers.

QD281.P6M375   547.84   59–14991 ‡

Library of Congress   ₁30₎

Note difference between
printed title and title for
catalog file

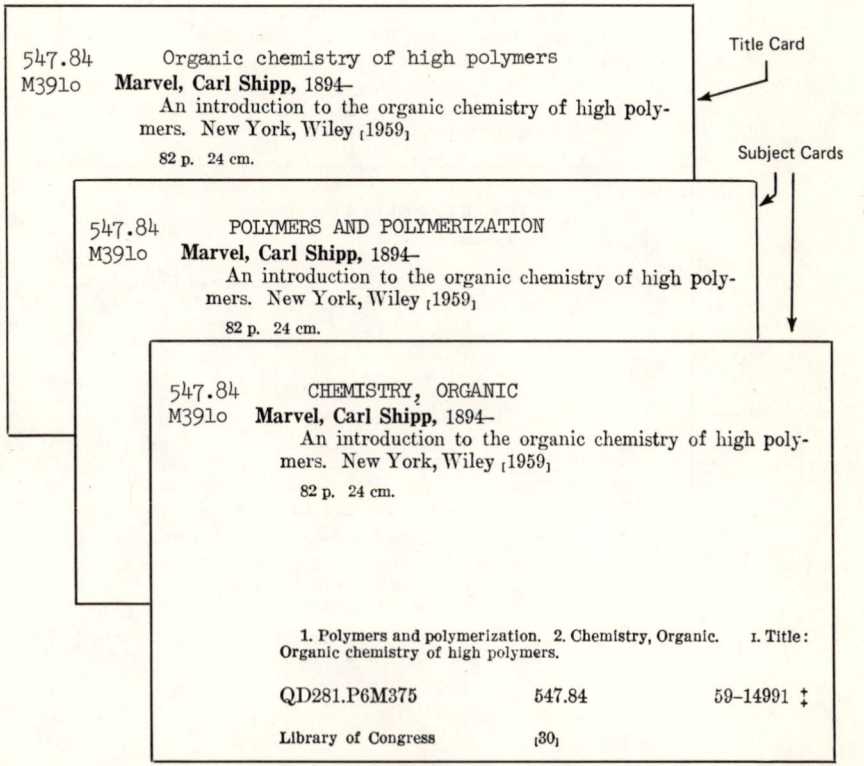

547.84   Organic chemistry of high polymers
M391o   **Marvel, Carl Shipp,** 1894–
    An introduction to the organic chemistry of high poly-
mers. New York, Wiley ₁1959₎

     82 p. 24 cm.

Title Card

Subject Cards

547.84   POLYMERS AND POLYMERIZATION
M391o   **Marvel, Carl Shipp,** 1894–
    An introduction to the organic chemistry of high poly-
mers. New York, Wiley ₁1959₎

     82 p. 24 cm.

547.84   CHEMISTRY, ORGANIC
M391o   **Marvel, Carl Shipp,** 1894–
    An introduction to the organic chemistry of high poly-
mers. New York, Wiley ₁1959₎

     82 p. 24 cm.

1. Polymers and polymerization. 2. Chemistry, Organic. ɪ. Title:
Organic chemistry of high polymers.

QD281.P6M375   547.84   59–14991 ‡

Library of Congress   ₁30₎

Figure 3. Catalog cards, Dewey Decimal classification system

2. The subject card is made from an author card by typing or printing the subject in capital letters directly over the author's name.

3. A co-author (or co-editor) card is made from the author card by typing or printing the co-author's name in upper and lower case letters directly over the first author's name.

There is much useful information on the catalog card. If carefully read, it may save you the time of finding and scanning a book to discover that it does not meet your present need. Some of the information is for the librarian, but you will be interested in the following entries.

1. Author's full name, birth date, and date of death if he was not living when the card was printed.

2. Directly under the author's name is the title of the book as printed on the book and on the title page. If the book is available in a second or later edition, it will be so noted.

3. The name and location of the publisher and the publication date.

4. If it is a single volume book, the following line tells you the number of pages, whether or not it is illustrated, and the size. (If the book is of unusual size, it may be shelved in a place separate from other books on the same topic.)

5. If the book includes a bibliography, the card will tell you and may even designate the relevant page numbers.

6. A short paragraph may describe the content if the title is insufficiently descriptive to the master librarian who prepared the card.

7. The card will usually list subject headings, slightly different from the title, under which subject cards are prepared and filed. Subject headings are carefully chosen. Subjects such as Organic Chemistry, Physical Chemistry, and Analytical Chemistry would all be under Chemistry, each with a modifier, *i.e.,* Chemistry, Organic. This procedure groups cards on similar subjects in the catalog drawers.

Roman numerals following the subject refer to joint authors, sometimes trans-lators, etc., for whom additional catalog cards have been prepared. A Roman numeral that just says "Title" means the title card is headed by the exact title of the book, omitting only "a," "an," "the," etc. However, when other relatively unimportant words such as "An introduction to . . ." are omitted, the Roman numeral entry will say "Title:" and follow that by the exact words used on the heading of the title card. *See* Figure 3.

8. At the bottom of the card there are classification numbers for the book under both the Library of Congress and Dewey Decimal Number Classification Systems. One of these will be repeated in the call number given at the upper, left hand corner depending upon which system your library uses.

The information to go on a catalog card is determined by an expert librarian. The Card Division of the Library of Congress prepares a printed author card for each book appearing under a U.S. copyright. These may be purchased by other libraries that usually print or type their own title and subject cards from author cards.

Periodical Indexes

Finding information in books, pamphlets, encyclopedias, dictionaries, and handbooks through the catalog cards is relatively straightforward. However, periodicals that contain shorter articles on many subjects pose a different problem. Libraries include some of their periodicals in the catalog file, but the most convenient listing consists of single line entries that tell which periodicals are received and what time span is covered by the ones on hand. This listing gives no information about which issue or volume should be examined to find an article on a specific subject. This information must be gained from a periodical index.

Chemical Abstracts, one of the most useful periodical indexes, has already been mentioned and will be discussed in detail later. However there are thousands of articles in hundreds of magazines each year that may be of interest to you but are not abstracted by CA. How do you locate them?

The *Applied Science and Technology Index* is the most important such index for chemical technicians. It is published by the H. W. Wilson Co., 950 University Avenue, Bronx, New York 10452, and is described in Table III and Figure 4. Note that it is a subject index and, in addition, refers to each article by title, author, volume number, page, and date. Articles requiring more than one heading are listed under as many subjects as may be required. This index was started in 1958, replacing the older *Industrial Arts Index.*

The most widely used index to nontechnical literature is the *Readers' Guide to Periodical Literature,* also published by the H. W. Wilson Co. If you need a news article, a general report of some meeting, or a nontechnical description of a new discovery, this index could be helpful. Its references are indexed by both subject and author.

Any technician who intends to use a library properly should become acquainted with these two indexes. Special guides to using the indexes can be obtained by writing to the H. W. Wilson Co.

DRUGS

   Control of the mind. F. M. Berger. bibliog Am Scientist 55:67–71 Mr '67

   Fenamates in medicine. Manuf Chem 38:4 Mr '67

   Magnesium pemoline; failure to affect in vivo synthesis of brain RNA. N. R. Morris and others. Science 155:1125-6 Mr 3 '67

      See also

Drug trade

Analysis

   Review of analytical applications; pharmaceuticals and related drugs. J. G. Theivagt and others. Anal Chem 39 [no 5]: 191R-239R bibliog (p212R-39R) Ap '67

DRYING apparatus

   Air and gas filtration; new drying systems provide moisture-free air. S. Elonka. flow diag il diags Power 111:183-90 F '67

   Compressed air drying soaks up problem-causing moisture. A. K. Shenk. diag Plant Eng 21:128-9 F '67

   Fluid-bed drying. W. E. Clark. bibliog il Chem Eng 74:177-84 Mr 13 '67

   Granular material drier. Engineer 223:357 Mr 3 '67

   Regeneration cycle gives desiccants new life. H. J. Calon. diag Plant Eng 21:114-17 F '67

      See also

Electric ovens

Ovens

Figure 4. A sample entry in the Applied Science and Technology Index

Table III. A Sample List of Periodicals Indexed in
Applied Science & Technology Index

AIChE Journal	Electrochemical Technology
Adhesives Age	Food Technology
American Chemical Society, Journal	Franklin Institute Journal
American Oil Chemists' Society, Journal	Industrial Photography
Analytical Chemistry	Iron Age
Bell Laboratories Record	Journal of Applied Chemistry
Canadian Journal of Chemical Engineering	Lubrication Engineering
Chemical & Engineering News	Modern Plastics
Chemical & Process Engineering	Physics and Chemistry
Chemical Engineering	Rock Products
Chemical Engineering Progress	Safety Maintenance
Chemistry and Industry	Soap and Chemical Specialties

Starting a Search

Usually the simplest search is locating one or more specific facts about a particular substance or compound. Physical or chemical properties can frequently be found in a chemical handbook, dictionary, or the "Merck Index." If the chemical properties are not found there, try a chemical encyclopedia. The next step might be to find a set of collected tables. It is probably a waste of time to try to hunt for primary sources initially. For very rare substances, a search for specific facts may become a comprehensive detective job.

Comprehensive searches or literature surveys should be deliberately and carefully planned.

1. Make sure the assignment is thoroughly understood.

2. Be certain you understand the nomenclature used for the assignment. If necessary, start with a comprehensive English language dictionary. From there you should continue with general references such as chemical dictionaries and encyclopedias.

3. If the subject of interest is only a few months old, it may be appropriate to look at the latest *Applied Science and Technology Index.*

These steps are helpful in defining the shape and size of the problem and insuring maximum efficiency in the comprehensive search. Go from the general to the specific. You can now search the card catalog for books, consult *Chemical Abstracts,* or go to other indexes. Always remember the time lag between the date of publication of any information and its appearance in an abstract, index, or book. If the search is being made for a satisfactory synthetic method, an old method may be suitable and might be found in an old volume of "Organic Syntheses." If it is an official analytical procedure, you should look in the latest official publications of analytical procedures published by the organization involved. For example, for analysis of foods, see the *Official Methods of Analysis* of the Association of Official Analytical Chemists. (*See* Chapter 15 for information on so-called official analytical procedures.)

The National Referral Center in the Library of Congress publishes a "Directory of Information Resources" which lists organizations that provide information services or literature searches. It can be obtained from the Superintendent of Documents, Washington, D.C. 20402.

If statistical information is needed, it may be available in yearbooks, almanacs, trade journals, or government publications. Some sources of particular relevance to your job may be poorly indexed for your needs, so you may want to keep your own card index file under subject headings that might save you much time in later searches.

At this point the book, "Searching the Chemical Literature," ADVANCES IN CHEMISTRY SERIES, No. 30, is very useful. Each chapter discusses the techniques and problems of using a particular kind of source publication.

If you know of a reference not available in your library, ask your librarian about inter-library services. The reference usually can be borrowed from another library, or the pertinent portions can be obtained through inter-library copying services or on micro-film.

Chemical Abstracts

The most complete reporting of all publications in chemistry appears in *Chemical Abstracts.* This journal is published weekly and contains short summaries of U.S. and foreign books and articles. Usually the abstract appears from three months to one year after the original publication. *Chemical Abstracts* has been published by the American Chemical Society since 1907. Complete collections are found in large university and industrial chemical libraries.

A typical issue of the *Chemical Abstracts* contains summaries (abstracts) of about 6,000 articles. These are grouped by subject, such as Synthetic High Polymers, but are numbered in a series from 1, the first abstract printed in the first issue of the volume, to the highest number for the last abstract of the volume. In addition, a small letter follows the number. Before 1967 this letter indicated the position on the page where the abstract was located, but since then the letter is for computer coding purposes only.

A typical abstract begins as shown in Figure 5.

The numbers in the figure have the following significance:

1. *The abstract number.*
2. *Title.* An exact or shortened title of the article if it was in English or a literal translation of a foreign language title.
3. *Author names,* with the last name first.
4. *Address.* Location of the author's laboratory.

5. *Journal Title.* The abbreviation of the name of the publication in which the article originally appeared.
6. *Year* of the original publication.
7. *Volume* number and (issue number).
8. *Pages* on which the article is found.
9. *Language* in which the article was printed.

A similar format is used for abstracts of books or patents.

Using the Abstracts

Let's suppose you wish to check recent work done on the reaction of diborane with hydrocarbons. Go to the library and use the *Chemical Abstracts* to start your search.

First, check the Index issues of *Chemical Abstracts* for the last year or any previous years. Find the volume labeled Subject Index and look at the listings under Diborane. Figure 6 is a picture of a page showing these listings in the Subject Index. Look down the list and read the short subject descriptions, one of which says:

reaction of
  with methylcyclopropene **70**:106013s

If this reaction is of interest to you, find Volume 70 of *Chemical Abstracts,* which is the January–June period for

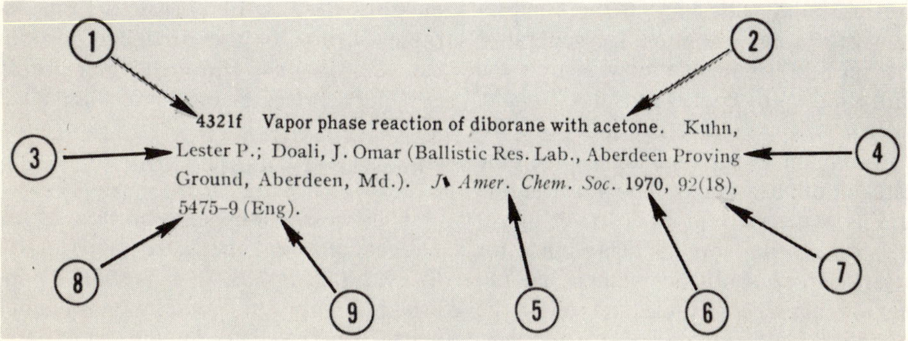

4321f Vapor phase reaction of diborane with acetone. Kuhn, Lester P.; Doali, J. Omar (Ballistic Res. Lab., Aberdeen Proving Ground, Aberdeen, Md.). *J. Amer. Chem. Soc.* 1970, 92(18), 5475-9 (Eng).

Figure 5. Information provided in an abstract from Chemical Abstracts

Diborane(6)
  70:115211g
  anti-Markovnikoff reaction of, with epoxides
      in presence of boron fluoride, 70:37344m
  beryllium complexes, 70:82604g
  from boron chloride, app. for, 70:P 79635n
  catalysts from hydrogen peroxide and, for
      graft polymn. of styrene on isoprene
      rubber, 70:P 58760j
  catalysts from nickel diisopropylsalicylate
      and, for polymn. of butadiene, 70:P
      29887a
  compd. with neopentanetetrayltetrakis[di-
      phenylphosphine] (1:1), structure of, 70:
      57951k
  in detn. of amines, 70:25497a
  diffusion in hydrogen and, 70:40886v
  doping of silicon by, 70:42154d
  with epoxides, mechanism of, 70:28437s
  expansion of, in propellant systems, 70:
      98394z
  manuf. of, 70:P 69700r
  methylcyclohexenyl acetate redn. by, 70:
      10774q
  mol. orbitals of, 70:60972y
  mol. structure of, 70:32435g
  photolysis of hexafluoroacetone and, 70:
      33212u
  pyroglutamic acid redn. by, in proteins, 70:
      111633y
  reaction of
      with hexamethylhexaazatetraphospha-
          adamantane, 70:10949a
      with Lewis bases, 70:78047k
      with methylcyclopropene, 70:106013s
      with sodium (dimethylaminato)trihydro-
          borate(1-), 70:96080v

Figure 6. A subject index listing in Chemical Abstracts

1969 and look in the issue which contains abstract number 106013s. This issue is the June 9, 1969 edition of *Chemical Abstracts,* and the cited abstract is shown in Figure 7.

The abstract may show that the work done is related to your own project, in which case you might choose to read the original article in *Angwandte Chemie.* On the other hand, you might not be interested and instead would continue

the search by going back to the Subject Index.

In a similar way you can start your search by looking up the work of an author, rather than a subject. Suppose you know that Dr. Roland Koester of the Max Planck Institute has been very active in the field of diborane chemistry. You can then consult the Author Index of *Chemical Abstracts* and note, as in Figure 8, that Koester has published several articles on boron compounds which were abstracted in 1969. Again, you can check those which appear to have subject notations of interest to you, find the abstract, and, if you wish to pursue it further, obtain the original article.

Other keys which may be used with *Chemical Abstracts* are the Formula Index and the Patent Index. If you are given a formula or a patent number, you can start your search in much the same way as for Subject or Author.

If you decide to obtain the original article after checking the abstract, you will not only be able to read the report of the work done, but you will also be able to check the references at the end of the article. This procedure can often lead to the information you want more directly than continuing the search in the *Chemical Abstracts.*

Computer Searches

The total number of publications of interest to chemists is increasing rapidly.

106013s trans-2-Methylcyclopropanol from 1-methylcyclopropene. Koester, Roland; Arora, S.; Binger, Paul (Max-Planck-Inst. Kohlenforsch., Muelheim/Ruhr, Ger.). *Angew. Chem., Int. Ed. Engl.* 1969, 8(3), 205 (Eng). Tris(*trans*-2-methylcyclopropyloxy)borane (I) is prepd. from the tris(cyclopropyl)borane and Me₃NO. The I soln. is heated to remove the (MeO)₃B/MeOH azeotrope and the title alc. (*trans*-II) contg. *cis*-II is obtained. N.M.R. spectral data are given.

Figure 7. An abstract from Chemical Abstracts

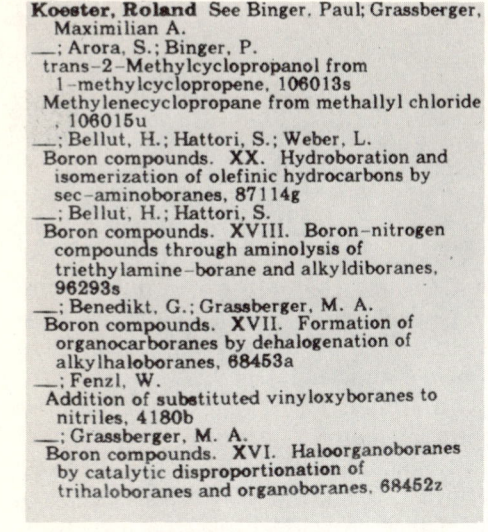

Koester, Roland See Binger, Paul; Grassberger, Maximilian A.
___; Arora, S.; Binger, P.
trans-2-Methylcyclopropanol from 1-methylcyclopropene, 106013s
Methylenecyclopropane from methallyl chloride, 106015u
___; Bellut, H.; Hattori, S.; Weber, L.
Boron compounds. **XX**. Hydroboration and isomerization of olefinic hydrocarbons by sec-aminoboranes, 87114g
___; Bellut, H.; Hattori, S.
Boron compounds. **XVIII**. Boron–nitrogen compounds through aminolysis of triethylamine–borane and alkyldiboranes, 96293s
___; Benedikt, G.; Grassberger, M. A.
Boron compounds. **XVII**. Formation of organocarboranes by dehalogenation of alkylhaloboranes, 68453a
___; Fenzl, W.
Addition of substituted vinyloxyboranes to nitriles, 4180b
___; Grassberger, M. A.
Boron compounds. **XVI**. Haloorganoboranes by catalytic disproportionation of trihaloboranes and organoboranes, 68452z

Figure 8. An author index listing in Chemical Abstracts

(*See* Figure 9.) The total number is now so large that a thorough search requires a great deal of time if it is done as we describe it here. Fortunately, the *Chemical Abstracts* published since 1966 have been entered into a computer which is programmed to perform searches (Figures 10 and 11). The computer can search in much the same way that you would do it yourself. Given information such as "Koester, Roland" as an author, or some key words such as "diborane," "synthesis," or "hydrocarbons" and instructions to search from 1969 to the current issue, the computer would retrieve and print the abstract numbers which relate to titles of papers which contain all of these words.

Selecting key words for a computer search requires careful thought. Remember that the computer has no capacity for judgment. For example, if the key words were "diborane," "synthesis," and "hydrocarbon," it could not select a title such as "Diborane synthesis of methylcyclopentane" because it does not rec-

ognize the relationship between "hydrocarbon" and "methylcyclopentane."

Most institutions contain a chemistry library supervised by a skilled librarian. If a computer search of *Chemical Abstracts* should be part of your work assignment, you should consult the librarian for help in developing the key word code as well as finding the method for submitting the information to the Columbus, Ohio, office of *Chemical Abstracts,* where the actual computer search is performed.

A computer literature search service is provided by the American Petroleum Institute in Washington. One use of this system is to make a continuous search of new publications. The computer prints a sheet showing all publications appearing in a stated period (a month, for example) containing references to key words selected by the person for whom the search is made. Figure 12 shows a section of the computer print made by the API service.

In this example, the person requesting the search is the supervisor of a large petrochemical laboratory and is responsible for determining small amounts of

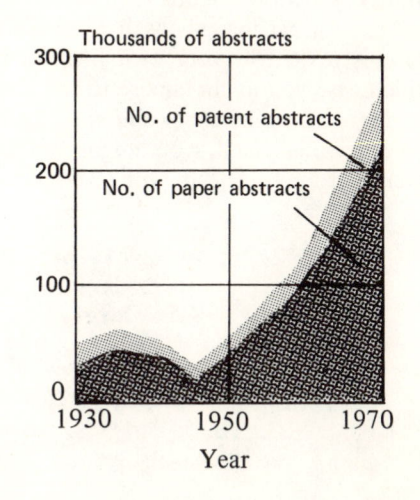

Figure 9. Growth of number of abstracts

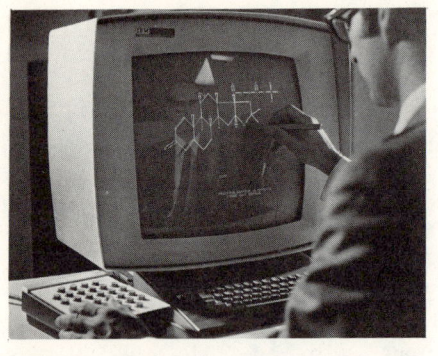

Figure 10. Entering structural formulas into the chemical registry system

Figure 11. Coding abstracts for computer processing

metals in petroleum products. He has sent in a set of key words for the computer program, some of which are shown in the top line of Figure 12. The first word—TRACE—is the most important; all titles must have this word in order to be selected by the computer. Since the laboratory analyses only small amounts of metals, they don't want to see any of the thousands of abstracts on semimicro or macro scale analysis. The words "determination" (abbreviated, DETN) and ANAL must also appear in the title. Using an abbreviated word means that the computer will accept "analysis," "analytical," or analyzer" as variations of ANAL. Finally, five metals of interest are listed as key words, any

of which will be selected by the computer. Therefore any publication which has TRACE in the title, plus DETN or DETERMINATION or ANAL, and the name of any of the five metals will be selected for printing by the computer.

The printed reference shows:

1. The key words found in the title.

2. The authors, laboratory, and address.

3. The title.

4. The publication name, volume, issue, year, and pages.

5. Key words used by the API in describing the publication.

Notice that these are words like WATER and COBALT and OSCILLOPOLAROG in the list which were not in Line 1.

```
① TRACE, DETN , ANAL , LEAD, CADMIUM, COPPER, MANGANESE, ZINC
  METAL

② ES SOUAB A   LAB. CHIM. ANAL., FAC. SCI. AGRON., GEMBLOUX, BEL
  NANGNIOT P

③ MEASUREMENT OF TRACES OF CADMIUM, COBALT, COPPER, IRON, MANGAN
  ESE, LEAD, AND ZINC IN A FEW BELGIAN MINERAL WATERS, BY DIFFER
  ENTIAL OSCILLOPOLAROGRAPHY AT AN APPLIED VOLTAGE.

④ CHIM. ANAL. (PARIS)
  VOLUME 0053, ISSUE 0003, YEAR 1971, PAGE 176-82

⑤ MANGANESE ** DETN ** MINERAL ** WATER ** CADMIUM ** COPPER **
  COBALT ** METAL ** ZINC ** LEAD ** IRON ** OSCILLOPOLAROG ** A
  NAL **                                                        $

⑥ IS THIS CITATION USEFUL?   YES    NO    CANNOT TELL    COMMENT
  AN 130224Q P 2249 EN 01 TW 000 WT 000 S C2474 TP ARTC L
```

Figure 12. An example of a computer search

6. Some questions designed to test whether the search service is really valuable to the user, and finally, an identifying series of code numbers and letters used by the search service.

References to Library Use

Cook, M., "The New Library Key," 2nd ed., H. W. Wilson, New York, 1963.

Crane, E. J., Patterson, A. M., Marr, E. B., "A Guide to Literature of Chemistry," 2nd ed., John Wiley and Sons, New York, 1957.

Fieser, Louis F., Fieser, Mary, "Style Guide for Chemists," Reinhold, New York, 1960.

Gensler, W. J., Gensler, K. D., "Writing Guide for Chemists," McGraw-Hill, New York, 1961.

Herner, Saul, "A Brief Guide to Sources of Scientific and Technical Information," Information Resources Press, Washington, D.C., 1969.

Mellon, M. G., "Chemical Publications," 4th ed., McGraw-Hill, New York, 1965.

Appendix

Appendix

Glossary

Accuracy—the degree to which a given answer agrees with the true value.

Activity—the effective concentration of ions in solution.

Activity coefficient—the ratio of activity to concentration.

Aliquot—a volume of a liquid which is a fractional part of a larger volume.

Allotropic forms—two or more forms of an element that exist in the same physical state (such as graphite and diamond).

Alpha particle (alpha radiation, alpha ray)—a positively charged particle (a helium nucleus) consisting of two neutrons and two protons.

Amphoteric—having ability to act as either an acid or a base.

Anion—a negatively charged ion.

Anode—the electrode at which oxidation occurs in electrochemical cells.

Anodizing—coating a metal surface with a metal oxide by anodic oxidation.

Atom—the smallest particle of an element retaining the properties of that element.

Atomic weight—the relative weight of an atom referred to carbon-12 which has a weight of 12.0000 amu.

Atmosphere—a unit of pressure which is equivalent to 760 torr.

Avogadro's Hypothesis—equal volumes of gases under the same conditions contain equal numbers of molecules.

Avogadro's Number—6.023×10^{23}, which is the number of units of any species in one mole.

Background radiation—radiation extraneous to an experiment.

Beer's Law—absorbance of a solution is proportional to the length of light path and to the concentration.

Beta particle (beta radiation, beta ray)—an electron which has been emitted by an atomic nucleus.

Capillary—a tube having a very small inside diameter.

Carbanion—an organic ion carrying a negative charge on a carbon atom.

Carbonium ion—an organic ion carrying a positive charge on a carbon atom.

Catalyst—a substance which changes the rate of a reaction but is not itself consumed.

Cathode—the electrode at which reduction occurs in electrochemical cells.

Cation—a positively charged ion.

Chromatography—a group of methods for separating substances by selective partitioning between two phases, one of which moves with respect to the other.

Cis-—prefix used to indicate that groups are located on the same side of a bond about which rotation is restricted.

Coefficient of expansion—the ratio of the change in length or volume of a body to the original length or volume for a unit change in temperature.

Conjugated—two double bonds separated by one single bond (—C=C—C=C—).

Corrosion—the slow conversion of a metal to an oxidized form.

Coulometry—the quantitative application of Faraday's Law to the analysis of materials. The current and the time are the usual variables measured.

Curie (Ci)—the basic unit used to describe the intensity of radioactivity in a sample of material. One curie equals 37 billion disintegrations per second or approximately the radioactivity of 1 gram of radium.

Current efficiency (%)—the percentage of the electrons passing to or from an electrode that reacts with the material in question.

Daughter—a nuclide formed by the radioactive decay of a different (parent) nuclide.

Decay (radioactive)—the change of one radioactive nuclide into a different nuclide by the spontaneous emission of alpha, beta, or gamma rays.

Denaturation—a process pertaining to a change in structure of a protein from a regular to an irregular arrangement of the poly-peptide chains.

Density—the mass per unit volume of a substance.

Derivative—a compound which can be imagined to arise from a parent compound by replacement of one atom with another atom or group of atoms.

Differential Scanning Calorimetry (DSC)—a technique for measuring the temperature, direction, and magnitude of thermal transitions in materials by heating or cooling a sample and comparing the amount of energy required to maintain its rate of temperature increase or decrease with an inert reference material under similar conditions.

Differential Thermal Analysis (DTA)—a technique for observing the temperature, direction, and magnitude of thermally induced transitions in a material by heating or cooling a sample and comparing its temperature with that of an inert reference material under similar conditions.

Distilland—the material in a distillation apparatus that is to be distilled.

Distillate—the material in a distillation apparatus that is collected in the receiver.

Dosimeter—a small, calibrated electroscope designed to detect and measure incident ionizing radiation.

Doublet—two peaks or bands of about equal intensity appearing close together on a spectrogram.

DP number—the "degree" of polymerization; more specifically, the average number of monomer units per polymer unit.

Electrolytic cell—an electrochemical cell in which chemical reactions are forced to occur by the application of an outside source of electrical energy.

Electromagnetic spectrum—the entire collection of radiant energy.

Electrophile—positively charged or electron deficient.

Electrophoresis—a technique for separating ions by rate and direction of migration in an electric field.

Electroplating—the deposition of a metal onto the surface of a material by an electrical current.

Eluate—the liquid obtained from a chromatographic column; same as effluent.

Emission spectra—a plot of energy emitted by excited atoms as a function of wavelength.

Enantiomer—one of the two mirror image forms of an optically active molecule.

Equilibrium—a state in which two opposing processes occur at the same rate in a closed system.

Essential oil—an extract of a plant which has a pleasant odor or flavor.

Film badge—a small patch of photographic film worn on clothing to detect and measure accumulated incident ionizing radiation.

"Fingerprint" spectra—infrared spectra used to identify unknown compounds.

Fission—the splitting of a heavy nucleus into two roughly equal parts (which are nuclei of lighter elements), accompanied by the release of a relatively large amount of energy in the forms of kinetic energy of the two parts and the emission of neutrons and gamma rays.

Free radical—a highly reactive chemical species carrying no charge and having a single unpaired electron in an orbital.

Fuel cell—a voltaic cell that converts the chemical energy of a fuel and an oxidizing agent directly into electrical energy on a continuous basis.

Functional groups—certain groups of atoms having the same properties when appearing in various molecules.

Gamma ray—a highly penetrating type of nuclear radiation similar to x-radiation, except that it comes from within the nucleus of an atom and, in general, has a higher energy.

Galvanizing—placing a thin layer of zinc on a ferrous material to protect the underlying surface from corrosion.

Geiger counter—a gas-filled tube which discharges electrically when ionizing radiation passes through it.

Gem-dimethyl group—two methyl groups on the same carbon atom.

Half-life—the time in which half the atoms of a particular radioactive nuclide disintegrate.

Heat of combustion—the quantity of heat released when 1 gram of a substance is oxidized.

Heat of fusion—the quantity of heat required to melt 1 gram of a solid substance.

Heat of vaporization—the quantity of heat required to vaporize 1 gram of a liquid substance.

Heavy metals—those metals which have ions that form an insoluble precipitate with sulfide ion.

Heterocycle—a compound containing a ring of atoms at least one of which is an atom other than carbon.

HETP—Height Equivalent to a Theoretical Plate. The column length divided by the number of equilibrium steps in the column.

Hybridization—the mixing of atomic orbitals in the bonding process to produce an equal number of molecular orbitals of identical character.

Hydrogen bond—a loose bond between two molecules one of which contains an active hydrogen; the other contains an electro-negative atom such as O, N, F, etc.

Hydrolysis—a reaction in which water molecules react with another species with the release of H^+ or OH^- ions from the water molecules.

Inert phase—a nonreactive packing material used in chromatography.

in situ—at the original location.

Ionizing radiation—radiation that is capable of producing ions either directly or indirectly.

Irradiate—to expose to some form of radiation.

Isoelectric point—the pH at which there is no migration in an electric field of dipolar ions.

Isotopes—atoms which have the same number of protons but a different number of neutrons in their nuclei.

Kinetic energy—energy caused by motion.

Kinetic molecular theory—the theory which states that all matter is composed of particles that are in constant motion.

Ligands (coordination groups)—the molecules or anions attached to the central metal ion.

Lipid—an ether or benzene soluble product occurring in nature, usually limited to oils, fats, waxes, and steroids.

Lyophilization—a process whereby the material is frozen, a vacuum applied, and the water and low boiling compounds removed by sublimation (freeze-drying).

MAC—maximum allowable concentration of a toxic substance under prescribed conditions in an atmosphere to be breathed by humans.

Manostat—a device for maintaining a constant pressure.

Mass—a measure of the amount of matter. The mass of an object is constant irrespective of its position in the universe.

Mass number—approximately the sum of the numbers of protons and neutrons found in the nucleus of an atom.

Meniscus—the curved upper surface of a liquid column.

Metallic bond—the type of bond found in metals as the result of attraction between positively charged ions and mobile electrons.

Metalloid—an element that exhibits properties intermediate between metals and nonmetals.

Metric system—a system of measurement in which the units are related by powers of ten.

Mobile phase—a liquid or gas that is used to move the sample during a chromatographic process.

Mole—6.023×10^{23} units.

Molality—the concentration of a solution expressed as moles of solute per 1,000 g of solvent.

Molarity—the concentration of a solution expressed as moles of solute per liter of solution.

Molecular distillation—the distillation of viscous materials under high vacuum so that the mean free path is longer than the distance from the material to the condenser.

Molecular orbital—overlapping of atomic orbitals or hybrid orbitals in covalent bonding.

Molecular weight—the relative weight of a molecule based on the standard carbon-12 = 12.0000.

Molecule—the smallest particle of a compound which retains the properties of that substance.

Monochromator—a device which separates light into its component wavelengths.

Monosaccharide—a small carbohydrate molecule which is the monomeric unit from which the polymeric carbohydrates are composed.

Mull—a suspension of a finely powdered sample in oil.

Mutarotation—the interconversion of two forms of a sugar molecule.

Neat liquid—pure liquids as opposed to solutions.

Neutron—an uncharged nuclear particle with a mass approximately equal to that of a proton.

Nonpolar molecule—a molecule in which the electrical charges are symmetrically distributed around the center.

Nuclear reactor—a device in which a nuclear chain reaction can be initiated, maintained, and controlled.

Nucleon—a constituent of the nucleus; that is, a proton or a neutron.

Nuclide—any species of atom that exists for a measurable length of time, distinguished by its atomic weight, atomic number, and energy state.

Optical activity—the property of a molecule involving rotation of plane polarized light.

Overtone band—a band which occurs at twice the frequency of the fundamental band, as in IR analysis.

Oxidation—a reaction with oxygen or oxygen compounds; a loss of electrons; a positive change in valence.

Oxidation number—an arbitrary number assigned to represent the number of electrons lost or gained by an atom in a compound or an ion.

Oxidizing agent—a substance that oxidizes another substance while it is reduced.

Packing—small particles used to fill a chromatographic column.

Parent—a radionuclide that decays to another nuclide which may be either radioactive or stable.

Parts per million (ppm)—an expression for concentration; one part of the substance per million total parts.

Peptide—two or more amino acids joined by peptide linkages (—CONHC—).

Peroxide—a molecule containing —O—O— bonds.

Pi-bond (π-bond)—covalent bond formed by overlapping of p atomic orbitals on adjacent atoms in a molecule.

Plasticizer—a liquid of low volatility (high boiling point) which is added to soften polymers.

Polar molecule—a molecule in which the electrical charges are not symmetrically distributed; the center of positive charge is separated from the center of negative charge.

Potentiostat—an instrument designed to maintain a constant potential between two electrodes in a solution.

Precision—a measure of the reproducibility of repetitive results.

Proton—the nucleus of a hydrogen atom having a mass of approximately one amu and carrying one unit of positive charge.

Pyrolysis—the breakdown of a material by heating, usually in the absence of oxygen.

Pyrophoric material—a substance which can heat up substantially or ignite spontaneously upon exposure to air.

Racemate—a mixture having no optical activity which consists of equal amounts of enantiomers.

Rad (Radiation Absorbed Dose)—The basic unit of absorbed dose of ionizing radiation.

Radioactive dating—a technique for estimating the age of an object by measuring the amounts of various radioisotopes compared with stable isotopes.

Radioactivity—the spontaneous decay or disintegration of an unstable atomic nucleus accompanied by the emission of radiation.

Reflux ratio—the ratio of the amount of material returned to a column compared with the amount collected per unit time.

Resonance—characterized by stability shown by a molecule, ion, or radical to which two or more structures differing only in the distribution of electrons can be assigned.

Resonance wavelength—the wavelength which corresponds to the energy required to shift a ground state electron to a higher energy level.

Reducing agent—a substance that reduces another while it is being oxidized.

Reduction—a removal of oxygen or an addition of hydrogen; a gain of electrons; a negative change in valence.

R$_f$ value—ratio of the distance a compound moved to the distance the solvent moved on a paper or thin layer chromatogram.

Scaler—an electronic instrument for counting radiation-induced pulses from radiation detectors such as a Geiger–Müller tube.

Scintillation counter—an instrument that detects and measures gamma radiation by counting the light flashes (scintillations) induced by the radiation.

Sigma bond (σ-bond)—covalent bond formed by overlapping of s—s, s—sp^3, s—sp^2, or s—sp orbitals on adjacent atoms in a molecule.

Significant figures—the figures of a number that are thought to be correct. Never include any figures beyond the first one in which there is some doubt.

Specific gravity—the ratio of the density of a substance to the density of a standard substance. For solids and liquids, water is the standard substance; for gases, air is the standard substance.

Specific heat—the number of calories required to raise the temperature of 1 gram of a substance by 1°C.

Spectrometer—an instrument which separates electromagnetic radiation into its component wavelengths and measures the intensity of the radiation at each wavelength.

Standard conditions—the set of conditions adopted for standard pressure and temperature (STP) which is 1 atmosphere and 0°C.

Standard electrode potential—the potential compared with a hydrogen electrode which exists when the electrode is immersed in a solution of its ions at unit activity.

Stationary phase—the immobile phase used in a chromatographic separation.

Stereoisomers—isomers having different arrangement of the atoms in space but the same number of atoms, kinds of atoms and sequence of bonding of the atoms.

Steric hindrance—a condition in which a reaction is slowed or stopped because the size of the groups near the reaction site blocks approach to the site.

Sublimation—change of state from a solid to a vapor without going through a liquid state.

Substance—matter of definite chemical composition (*i.e.*, water is 1 part hydrogen and 8 parts oxygen by weight).

Thermogravimetric analysis (TGA)—an analytical technique in which the weight of a sample is measured continuously as its temperature is increased or decreased.

Threshold Limit Value (TLV)—an established upper boundary limit of concentration of a toxic substance in an environment which, if exceeded, may cause bodily harm.

Tracer—a small amount of radioactive isotope introduced into a system in order to follow the behavior of some component of that system.

Trans- —prefix used to indicate that groups are located on the opposite sides of a bond about which rotation is restricted.

Valence—combining capacity of an element, *i.e.*, valence of O in H_2O is 2—.

Vapor pressure—the pressure exerted by the gaseous molecules which exist in the space over a liquid.

Viscosity—the internal resistance of a fluid to flow.

Visible spectrum—light in the wavelength range of about 4,000–6,500 A.

Voltaic cell—an electrochemical cell in which an electrical current is generated by a chemical reaction.

Weight—a measure of the gravitational force acting on an object with a given mass.

X-ray—high frequency electromagnetic radiation lying between gamma rays and short ultraviolet rays in the energy spectrum.

Zwitterion—an ion which contains both a negative and a positive charge, same as a dipolar ion.

Abbreviations

Units of Measurement

A—ampere
atm—atmosphere
amu—(unified) atomic mass unit

cal—calorie, thermochemical
cm—centimeter
Hz, cps—cycles per second

eV—electronvolt
esu—electrostatic unit
g—gram
g-atom—gram atom
Hz—hertz
in—inch
J—joule
kcal—kilocalorie
kg—kilogram
l.—liter
m—meter
μA—microampere
μl—microliter
μ—micron
mg—milligram

ml—milliliter
mm—millimeter
mμ—millimicron
mol—mole
m—molal (concentration)
M (M in biochemistry)—molar (concentration)
N (N in biochemistry)—normal (concentration)
lb—pound
rpm—revolutions per minute
sec—second
cm^2—square centimeter
m^2—square meter
V—volt
W—watt

Words Other Than Units

abs—absolute
anhyd—anhydrous
ca—approximate
aq—aqueous
at. wt.—atomic weight
av—average
bp—boiling point
calcd—calculated
CP—chemically pure
coeff—coefficient
compd—compound
concd—concentrated
concn—concentration
cf.—compare
const—constant
cor—corrected
cryst—crystalline
cd—current density
dec—decompose
diam—diameter
dil—dilute
dc—direct current
dist—distilled
eq—equation
equiv wt—equivalent weight
expt—experiment
exptl—experimental
fp—freezing point
(g), as in $H_2O(g)$—gas
ir—infrared
i. d.—inside diameter
insol—insoluble
i—iso-
(l), as in $H_2O(l)$—liquid
max—maximum

mp—melting point
m-—meta
min—minimum
mol wt—molecular weight
mol %—mole percent
neut equiv—neutralization equivalent
n—normal (chain)
nmr—nuclear magnetic resonance
obsd—observed
o-—ortho
o. d.—outside diameter
p(pp)—page(s)
p—para
ppm—parts per million
ppb—parts per billion (parts per 10^9)
ppt—precipitate
prepn—preparation
ref—reference(s)
s- (sec-)—secondary
(s), as in AgCl(s)—solid
sol—soluble
soln—solution
sp—specific
sp gr—specific gravity
std—standard
temp—temperature
t- (tert-)—tertiary
tlc—thin-layer chromatography
uv—ultraviolet
vac—vacuum
vp—vapor pressure
vol—volume
vol %—volume percent
wt—weight
wt %—weight percent

Index

Index

PERIODIC TABLE OF THE ELEMENTS

PERIOD / FAMILY	IA	IIA	IIIB	IVB	VB	VIB	VIIB	VIIIB			IB	IIB	IIIA	IVA	VA	VIA	VIIA	NOBLE GASES
1	1 H																	2 He
2	3 Li	4 Be											5 B	6 C	7 N	8 O	9 F	10 Ne
3	11 Na	12 Mg											13 Al	14 Si	15 P	16 S	17 Cl	18 Ar
4	19 K	20 Ca	21 Sc	22 Ti	23 V	24 Cr	25 Mn	26 Fe	27 Co	28 Ni	29 Cu	30 Zn	31 Ga	32 Ge	33 As	34 Se	35 Br	36 Kr
5	37 Rb	38 Sr	39 Y	40 Zr	41 Nb	42 Mo	43 Tc	44 Ru	45 Rh	46 Pd	47 Ag	48 Cd	49 In	50 Sn	51 Sb	52 Te	53 I	54 Xe
6	55 Cs	56 Ba	57 La	72 Hf	73 Ta	74 W	75 Re	76 Os	77 Ir	78 Pt	79 Au	80 Hg	81 Tl	82 Pb	83 Bi	84 Po	85 At	86 Rn
7	87 Fr	88 Ra	89 Ac	104	105	106	107											

6

58 Ce	59 Pr	60 Nd	61 Pm	62 Sm	63 Eu	64 Gd	65 Tb	66 Dy	67 Ho	68 Er	69 Tm	70 Yb	71 Lu

7

90 Th	91 Pa	92 U	93 Np	94 Pu	95 Am	96 Cm	97 Bk	98 Cf	99 Es	100 Fm	101 Md	102 No	103 Lw